普通高等教育“十一五”国家级规划教材

全国高等医药院校药学类专业第二轮实验双语教材

分子与细胞生物学实验与指导

（第2版）

主　　编　谭树华

副主编　郭　薇

编　　者（以姓氏笔画为序）

田　浤　朱娟娟　邱　郑

张　娟　陈　松　周　君

郑　珩　郭　薇　谭树华

中国健康传媒集团

中国医药科技出版社

内容提要

本教材是"全国高等医药院校药学类专业第二轮实验双语教材"之一。教材主要内容包括最基本、最常用的分子与细胞生物学实验技术，不仅对实验步骤做了详尽的描述，而且对相关理论以及概念进行了简要阐述，有利于学生进一步理解、掌握理论课上学习的概念。本教材为书网融合教材，即纸质教材有机融合电子教材、教学配套资源（PPT、微课、视频、图片等）、题库系统、数字化教学服务（在线教学、在线作业、在线考试），使教学资源更加多样化、立体化。

本教材主要适用于生物技术专业以及药学类专业的高年级本科生和低年级研究生，还可供从事分子与细胞生物学研究及其相关专业人员参考。

图书在版编目（CIP）数据

分子与细胞生物学实验与指导／谭树华主编．—2版．—北京：中国医药科技出版社，2019.12

全国高等医药院校药学类专业第二轮实验双语教材

ISBN 978-7-5214-1389-2

Ⅰ.①分…　Ⅱ.①谭…　Ⅲ.①分子生物学-实验-双语教学-医学院校-教学参考资料
Ⅳ.①Q7-33

中国版本图书馆CIP数据核字（2020）第000712号

美术编辑　陈君杞
版式设计　南博文化

出版　**中国健康传媒集团**｜中国医药科技出版社
地址　北京市海淀区文慧园北路甲22号
邮编　100082
电话　发行：010-62227427　邮购：010-62236938
网址　www.cmstp.com
规格　889×1194mm 1/16
印张　10 1/4
字数　250千字
初版　2003年8月第1版
版次　2019年12月第2版
印次　2019年12月第1次印刷
印刷　三河市万龙印装有限公司
经销　全国各地新华书店
书号　ISBN 978-7-5214-1389-2
定价　30.00元

版权所有　盗版必究
举报电话：010-62228771
本社图书如存在印装质量问题请与本社联系调换

获取新书信息、投稿、为图书纠错，请扫码联系我们。

序

Preface

教学是学校人才培养的中心环节，实验教学是这一环节的重要组成部分。“全国高等医药院校药学类专业实验双语教材”是中国药科大学坚持药学实践教学改革，突出提高学生动手能力、创新思维，通过承担教育部“世行贷款21世纪初高等教育教学改革项目”等多项教改课题，逐步建设完善的一套与药学各专业学科理论课程紧密结合的高水平双语实验教材。

本轮修订，适逢“全国高等医药院校药学类专业第五轮规划教材”及《中国药典》（2020年版）、新版《国家执业药师资格考试大纲》出版，整套教材的修订强调了与新版理论教材知识的结合，与《中国药典》（2020年版）等新颁布的法典法规结合。为更好地服务于新时期高等院校药学教育与人才培养的需要，在上一版的基础上，进一步体现了各门实验课程自身独立性、系统性和科学性，又充分考虑到各门实验课程之间的联系与衔接，主要突出了以下特点。

1. 适应医药行业对人才的要求，体现行业特色，契合新时期药学人才需求的变化，使修订后的教材符合《中国药典》（2020年版）等国家标准及新版《国家执业药师资格考试大纲》等行业最新要求。

2. 更新完善内容，打造教材精品。在上版教材基础上进一步优化、精炼和充实内容。紧密结合“全国高等医药院校药学类专业第五轮规划教材”，强调与实际需求相结合，进一步提高教材质量。

3. 为适应信息化教学的需要，本轮教材全部打造成为书网融合教材，即纸质教材与数字教材、配套教学资源、题库系统、数字化教学服务有机融合，为读者提供全免费增值服务。

4. 坚持双语体系，强调素质培养教材以实践教学为突破口，采用双语体系编写有利于加快药学教育国际接轨，提高学生的科技英语水平，进一步提升学生整体素质。

“全国高等医药院校药学类专业第二轮实验双语教材”历经15年4次建设，在各个时期广大编者的努力下，在广大使用教材师生的支持下日臻完善。本轮教材的出版，必将对推动新时期我国高等药学教育的发展产生积极而深远的影响。希望广大师生在教学实践中对本套教材提出宝贵意见，以便今后进一步修订完善，共同打造精品教材。

吴晓明

全国高等医药院校药学类专业第五轮规划教材常务编委会主任委员

2019年10月

前言

Preface

分子与细胞生物学迅猛发展，它在现代药学研究中发挥越来越重要的作用。分子与细胞生物学也是一门实验性很强的学科，因此，越来越多的学生希望通过实验课程更好地掌握分子与细胞生物学实验技术。为此，我们在原有初版教材基础上增补了细胞生物学实验部分，编写了本版《分子与细胞生物学实验与指导》。

本教材的一个主要特点是它不仅对实验步骤做了详尽的描述，而且对相关理论以及概念进行了简短阐述，有利于学生进一步理解掌握理论课上所学习的概念。本教材为书网融合教材，即纸质教材有机融合电子教材、教学配套资源（PPT、微课、视频、图片等）、题库系统、数字化教学服务（在线教学、在线作业、在线考试），使教学资源更加多样化、立体化。本教材配套的视频，对实验的基本操作步骤进行了演示，这将对学生正确了解基本的实验方法有所帮助。

本教材的内容主要包括最基本、最常用的分子与细胞生物学实验技术，适用于生物技术专业以及药学专业的高年级本科生和低年级研究生。

在此谨向中国药科大学生命科学与技术学院参与组织与校对书稿的研究生们表示感谢，此外特别感谢童守芳、陈雨婷和张晨曦同学，他们在视频制作过程中给予了很大帮助。

由于编写时间及水平所限，书中疏漏及不足之处，敬请广大师生批评指正，以期不断改善提高。

编　者

2019 年 11 月

目录

Contents

实验一　大肠埃希菌染色体 DNA 的分离与纯化

扫码“学一学”

扫码“看一看”

【导言】

从细胞中分离染色体 DNA 的方法有多种，目的都是为了获得高分子质量的 DNA 材料。染色体 DNA 一般很长，如大肠埃希菌的染色体有 4.7×10^3kb，长约 1mm。由于 B 型双螺旋 DNA 的宽度只有 1nm，DNA 分子对剪切力十分敏感，在分离过程中应尽量避免过度搅拌。为了进一步保护 DNA 样品的完整性，通常需要在缓冲液中添加乙二胺四乙酸（EDTA）以螯合 Mg^{2+}，这样可减少脱氧核糖核酸酶（DNase）的活性，因为该酶必须在有 Mg^{2+} 存在时才会起作用。

【目的】

掌握从大肠埃希菌细胞中分离纯化染色体的正确实验方法。

【原理】

本实验中添加去垢剂 sarcosyl 和十二烷基磺酸钠（SDS）是为了使蛋白变性并溶解细胞膜中的脂质从而导致细胞裂解，细胞裂解物常用 RNA 酶和蛋白进行水解处理。随后用酚：氯仿进行抽提，以去除残留的蛋白质，这时蛋白质进入有机相，如已变性可呈现在有机相与水相之间。乙醚处理是为了去除苯酚，而乙醇沉淀处理则可以浓缩 DNA，同时去除核苷酸、氨基酸以及低分子质量的寡核苷酸和肽。

本实验所采用的方法已成功地应用于大肠埃希菌和其他几种革兰阴性细菌（包括 *Rhizobia*，*Seeratia* 和 *Vibrio* sp.）的染色体 DNA 的抽提，而从其他类型的微生物（如*Bacillus* sp. 和酵母）中分离染色体 DNA 的方法则有所不同，主要区别在于裂解步骤不同。应用本实验所介绍的方法分离得到的染色体 DNA 样品易被限制酶切割，因此较适用于 Southern 杂交的研究。如果所制备的染色体 DNA 是用于构建克隆文库，要求得到较大分子质量的 DNA，则必须省略其中的振荡步骤。

【试剂与器材】

1. 大肠埃希菌 LE932 lacY galK2 galT22 metBl trpR55 supE44 supF58 hsdR514（r_k^-，m_k^+），LB 培养基培养至对数生长期末期（测定 OD_{600}，以估算每毫升培养液所含的细胞菌体数）。
2. 盛冰的容器。
3. TNE 缓冲液（10mM Tris－HCl，pH 8.0，10mM NaCl，0.1mM EDTA）。
4. TE 缓冲液（10mM Tris－HCl，pH 8.0，0.1mM EDTA）。
5. THE 缓冲液（50mM Tris－HCl，pH 8.0，20mM EDTA）。
6. 灭菌的微量离心管。

7. 灭菌的移液器枪头。

8. 灭菌的移液器。

9. RNA 酶（胰 RNA 酶 A，浓度为 10mg/ml，溶于 TE 缓冲液中，80℃预热 10 分钟使其中的 DNA 酶失活）置于冰浴中。

10. 链霉蛋白酶（浓度为 10mg/ml，溶于 TNE 缓冲液中，37℃预热 15 分钟，消化所有存在的 DNA 酶）置于冰浴中。

11. 2% *N* – sarcosyl（*N* – 月桂醇肌氨酸）溶于 HTE 缓冲液中。

12. 苯酚∶氯仿（1∶1，以 0.5M Tris – HCl，pH 8.0 平衡）；该氯仿为氯仿和异戊醇（24∶1*V*/*V*）的混合物。

13. 3.0M 醋酸钠。

14. 50℃水浴。

15. 37℃水浴。

16. 异丙醇（储存于 – 20℃）。

17. 70% 乙醇（室温）。

18. 乙醚（通风橱内）。

19. 聚乙二醇（PEG）400，用水按照 1∶1 的比例稀释，用于冲洗苯酚烧伤。

注意：苯酚可以造成严重的皮肤烧伤。在使用苯酚时必须戴手套，穿实验服，带防护眼镜，如果有可能带防护面罩。实验室必须备有 PEG 400 的水溶液（与水 1∶1 进行稀释），该溶液被推荐用来从体表除去苯酚。

20. 旋涡混合器。

21. 干冰 – 乙醇浴。

22. 真空干燥器。

【实验步骤】

1. 将 1.2ml 菌液置于微量离心管内，8000 转/分钟离心 15 秒。

2. 将菌体悬浮于 0.31ml HTE 缓冲液中，确保没有菌块。

3. 在 HTE 缓冲液中加入 0.35ml 2% *N* – sarcosyl。将离心管颠倒数次，充分混匀。

4. 加入 5μl RNA 酶，37℃水浴保温 15 分钟。再加入 35μl 链霉蛋白酶并在 50℃水浴加热直至细胞裂解完全（30 ~ 90 分钟）。

5. 旋涡振荡细胞裂解液 2 分钟以剪切 DNA。这一步骤可使 DNA 分子质量下降到 20kb 左右，还会降低裂解液的黏度，使之更易于用苯酚∶氯仿混合物及乙醚进行抽提。如果须保持大分子质量质粒的环状结构，就不要进行旋涡振荡。

6. 苯酚∶氯仿抽提：加入等体积（0.70ml）的苯酚∶氯仿（1∶1）的混合物，短暂（数秒）旋涡振荡混合，在微量离心机中最大速度离心 3 分钟使其分层，用灭菌的移液器仔细将顶层转移至一个新的微量离心管中，避免将白色絮状的中间层吸入。用苯酚∶氯仿对含 DNA 的水层再进行两次抽提。将苯酚∶氯仿倒入指定的废液回收容器内。

注意：苯酚和氯仿可以抽提脂类并且使蛋白变性，而异戊醇则可降低抽提时泡沫的产生并促进水相和有机相的分层。重蒸酚必须储存于 – 20℃，以防这些污染物的积累造成 RNA 和 DNA 的断裂与交联。如果苯酚带有任何颜色，那么在用于核酸纯化前必须对其进行重新蒸馏。苯酚或苯酚∶氯仿溶液用于抽提操作前，必须与缓冲溶液进行混合，以使其下层（有机相）水饱和，这样使用时对含有所要提纯的核酸的水相就不会产生任何吸收。

7. 等体积（0.70ml）乙醚抽提（通风橱内操作）：加入乙醚，短暂旋涡振荡，然后将顶层溶液弃至废液缸内。

注意：乙醚用于抽提水层中残留的痕量苯酚和氯仿。乙醚极易挥发和燃烧，应在化学通风橱内进行操作。

8. 沉淀 DNA：加入醋酸钠溶液，使其终浓度达到 0.3M（70μl），混合均匀，加入等体积的异丙醇（0.7ml）。加入异丙醇后，将离心管来回颠倒 5 ~6 次。在乙醇 - 干冰浴中冷冻 10 分钟。离心 5 分钟或更长时间并倾去上清。

注意：如果回收 DNA 时任存在问题，可以尝试延长离心时间。

9. 漂洗 DNA 沉淀：加入 70% 乙醇约 1.0ml，稍加混匀，然后离心 5 分钟，倾去上清。重复两次。可以除去残存的盐分以及残留的苯酚和氯仿。用封口膜将离心管口封好，并用针在上面扎几个眼，真空干燥以除去乙醇。

10. 将 DNA 沉淀溶解于 50μl 的 TE 缓冲液中。在管上做好标记，-20℃保存。

Experiment 1　Isolation and purification of *E. coli* chromosomal DNA

Introduction

There are several methods for isolating chromosomal DNA from cells. All have the common goal of obtaining high - molecular - weight material. Chromosomal DNA is quite long. For example, the chromosome of *E. coli* is 4.7×10^3 kb, which is about 1mm long. Because the double helix of DNA in the B form is only 1nm wide, DNA is extremely susceptible to shearing forces. In any isolation procedure then, excessively agitation should be avoided. As a further precaution, ethyenediaminetetraacetic acid (EDTA) is included in the buffers to chelate Mg^{2+}. This will reduce deoxyribonuclease (DNase) activity because of the Mg^{2+} requirement of these enzymes.

Purpose

Master the correct method for isolating and purifying chromosomal DNA from *E. coli* cell.

Principle

In the procedure described here, the detergents sarcosyl and sodium dodecyl sulfate (SDS) are added to denature proteins and solubilize lipids in membranes leading to cell lysis. The cell lysate is often treated with enzymes that hydrolyze RNA and proteins. This is generally followed by phenol of phenol : chloroform extraction to remove the remaining proteins, which will either enter the organic phase or, if denatured, appear at the interphase. Ether treatment is used to remove the phenol, and alcohol precipitation concentrates the DNA while removing nucleotides, amino acids, and low - molecular - weight oligonucleotides and peptides.

This procedure has been used successfully with *E. coli* and with several other types of Gram -

negative bacteria, including *Rhizobia*, *Seeratia* and *Vibrio* sp. Chromosomal DNA isolation procedures for other microorganisms differs mainly in the lysis step, such as for *Bacillus* sp. and for yeast. The protocol described here result in DNA that is easily cut with restriction enzymes and, therefore, is useful for Southern Hybridization studies. If the DNA is to be used to construct a clone bank, higher molecular weight DNA is desirable, so the vortex step should be eliminated.

Reagents and apparatus

1. *E. coil* strain LE932 lacY galK2 galT22 metB1 trpR55 supE44 supF58 hsdR514(r_k^-, m_k^+), grown in LB medium to late log phase(measure OD_{600} of culture in order to estimate the number of cell/ml).

2. Container of ice.

3. TNE buffer(10mM Tris - HCl, pH 8. 0, 10mM NaCl, 0. 1mM EDTA).

4. TE buffer(10mM Tris - HCl, pH 8. 0, 0. 1mM EDTA).

5. THE buffer(50mM Tris - HCl, pH 8. 0, 20mM EDTA).

6. Sterile microcentrifuge tubes.

7. Sterile pipet tips.

8. Sterile Pasteur pipets.

9. RNase on ice(pancreatic RNase A, 10mg/ml, in TE buffer, preheated to 80℃ for 10minutes to inactive DNase).

10. Pronase on ice (10mg/ml in TNE buffer preheated to 37℃ for 15minutes to digest any DNase present).

11. 2% *N* - sarcosyl in HTE buffer.

12. Phenol : chloroform(1 : 1, equilibrated with 0. 5M Tris - HCl, pH 8. 0), the chloroform is a mixture of chloroform and isoamyl alcohol(24 : 1 *V/V*).

13. 3. 0M sodium acetate.

14. Water bath set at 50℃.

15. Water bath set at 37℃.

16. Isopropanol(at -20℃).

17. 70% ethanol(at room temperature).

18. Ether(in hood).

19. Polyethylene glycol(PEG) 400 diluted 1 : 1 with water, to be used for washing phenol bums.

Note: Phenol can cause severe skin burns. Wear gloves, a labcoat, and safety glasses when handing it. Work in a fume hood if possible. The laboratory should contain a solution of PEG 400(diluted 1 : 1 with water); this is the solvent of preference for removing phenol from body surfaces.

20. Vortex mixer.

21. Dry ice - ethanol bath.

22. Vacuum dryer apparatus.

Procedures

1. centrifuge 1. 2ml cells in microcentrifuge at 8000rpm for 15 seconds.

2. Resuspend cell pellet in 0. 31ml HTE buffer making sure there are no cell clumps.

3. Add 0. 35ml 2% Sarcosyl in HTE buffer. Mix well by inversion of tube.

4. Add 5μl RNase, and incubate at 37℃ for 15minutes. Add 35μl of pronase and heat at 50℃ until lysis is complete(30 – 90minutes).

5. Vortex the lysate for 2minutes to shear the DNA. This step will lower the molecular weight of DNA molecules to around 20kb, and, because of the decreased viscosity, it will be easier to extract the lysate with penol: chloroform and ether. If you want to preserve the circularity of high – molecular – weight plasmids, do not vortex.

6. Phenol : chloroform extraction: Add an equal volume (0. 70ml) of phenol : choroform (1 : 1), vortex briefly (few seconds) to mix, centrifuge at full speed in a microcentrifuge for three minutes to separate the phase, and carefully remove top layer with a sterile Pasteur and add to a new microcentrifuge tube. Be sure not to take along the white flocculent interphase. Re – extract the aqueous DNA layer with phenol : chlorform two more times. Dispose of the phenol; chloroform in an approved waste receptacle.

Note: Phenol and chloroform will extract lipids and denature proteins, while isoamyl alcohol will reduces foaming during the extrcation and facilitates the separation of the aqueous and organic phase. Redistilled phenol should be stored at – 20℃ to prevent the accumulation of contaminants that cause breakdown and crossilinking of RNA and DNA. If phenol has any color, it should be redistilled before use in nucleic acid purification. Phenol or phenol: choloroform solutions should be mixed with a buffer solution prior to use in extraction procedures. The lower (organic) phase is water – saturated and will, therefore, not absorb part or all of the aqueous phase containing the nucleic acids that you are purifying.

7. Extract with an equal volume (0. 7ml) of ether (in hood): add ether, vortex briefly, and discard top layer in waste receptacle.

Note: Ether is used to extract traces of phenol and chloroform from the aqueous layer. Ether is highly volatile and extremely flammable; it should be used in a chemical hood.

8. Precipitate DNA: Add sodium acetate to 0. 3M (70μl), mix well, and add an equal volume of isopropanol (0. 7ml). After adding the isopropanol, invert tube five or six times. Freeze in an ethanol – dry ice bath for 10minutes. Centrifuge for 5minutes or longer and pour off supernatant.

Note: If you have trouble with recoveries of DNA, try increasing the length of time you centrifuge.

9. Wash DNA pellet: Add about 1. 0ml of 70% ethanol, vortex briefly, centrifuge for 5minutes, and pour off supernatant. Repeat two times. This removes residual salts as well as remaining phenol or chloroform. Wrap top of tube with Parafilm, poke holes in Parafilm with needle, then dry pellet under vacuum to remove ethanol.

10. Dissolve the precipitated DNA in 50μl of TE buffer. Label tube and store it at – 20℃.

扫码“学一学”

扫码“看一看”

实验二 利用紫外光谱测定 DNA 的浓度和纯度

【导言】

对 DNA 样品的浓度及相对纯度进行精确定量在分子操作过程中非常重要，无论是将纯化的 DNA 片断连接到克隆载体中，还是对小量制备的质粒或 PCR 扩增得到的 DNA 进行测序或进行其他更专业的操作，它在很大程度上影响实验结果及结果的重现性。

UV 分光光度法是一种精确、快速且无破坏性的测定 DNA 浓度的方法，它所测得的 DNA 的最低浓度可达 2.5μg/ml，因此在分子生物学实验中经常应用。

【目的】

通过本实验掌握利用紫外光谱测定 DNA 的浓度和纯度方法。

【原理】

DNA（或 RNA）分子中碱基的紫外吸收图谱在 260nm 处有一个特征吸收峰。这种吸收和系统中 DNA（或 RNA）量成正比。应用 1cm 光路，DNA 在 260nm 波长处的消光系数为 20。基于此消光系数，1cm 石英比色皿中，50μg/ml 双链 DNA 溶液在 260nm 处吸收值为 1。

DNA 双螺旋和单链分子之间的相互转换会使它的吸收水平产生一定的变化，但是这种偏差可以用一个特定的公式校正。这种方法方便且相对准确。

蛋白质的最大吸收大约在 280nm 处，这主要归因于色氨酸残基的吸收。因此，A_{260}/A_{280} 的比值可用来衡量 DNA 制剂的纯度，并且比值应在 1.65～1.85 之间。较低的比值说明了 DNA 制剂被蛋白质污染。由于苯酚在 270nm 处有最大吸收，如果 DNA 制剂被苯酚污染，则 A_{260} 将会异常高，导致 DNA 浓度被高估。

【试剂与器材】

1. DNA 样品。

2. 彼此配对的 1cm 石英比色皿。

注意：玻璃或塑料比色皿会吸收紫外光，因此，衡量 DNA 和 RNA 浓度时必须使用石英比色皿。

3. Eppendorf 管（微量离心管）。

4. Eppendorf 离心管架。

5. 紫外分光光度计（紫外灯已预热）。

【实验步骤】

1. 用 1∶20 或者更高的比例将样品 DNA 溶液在 TE（或 ddH_2O）中稀释。用 TE（或

ddH_2O）在紫外分光光度计波长260nm、280nm和310nm处做空白校正。

2. 将各DNA稀释溶液装入比色皿比色，读取上述3个波长下的光密度（*OD*）。

3. 记录光密度数据，通过计算确定DNA的浓度和纯度。

（1）对于单链DNA分子，其浓度（μg/ml）可以通过以下公式导出：

$$ssDNA = 33(OD_{260} - OD_{310}) \times \text{稀释倍数}$$

（2）对于双链DNA分子，其浓度（μg/ml）可以通过以下公式导出：

$$dsDNA = 50(OD_{260} - OD_{310}) \times \text{稀释倍数}$$

（3）对于单链RNA分子，其浓度（μg/ml）可以通过以下公式导出：

$$ssRNA = 40(OD_{260} - OD_{310}) \times \text{稀释倍数}$$

【注释】

1. OD_{310}是背景吸收，盐浓度高，该值就高。

2. DNA的OD_{260}：OD_{280}值应为1.8左右，表示在当前DNA样品中存在的杂质RNA和杂蛋白的量。

3. RNA的OD_{260}：OD_{280}值应为2.0左右。

Experiment 2　Determination of the concentration and purity of DNA by UV spectroscopy

Introduction

Accurate quantitation of concentration and relative purity DNA sample is very important in the molecular manipulation, because it can often significantly improve both the quality and consistency of your experimental results whether they involve the ligation of purified DNA fragments into cloning vectors, the sequencing of miniprep or PCR - generated DNA, or other more specialised procedures.

UV spectrophotometric determination is an accurate, rapid, and nondestructive method to determined concentrations as low as 2.5μg/ml of DNA and commonly used in molecular biology experiment.

Purpose

By the end of this laboratory exercise, you should be able to determine the concentration and purity of DNA by UV Spectroscopy.

Principle

The bases on DNA (or RNA) molecules have a characteristic UV absorption pattern with a peak at 260nm. This absorption is proportional to the amount of DNA (or RNA) in the system. Using a

1 - cm light path, the extinction coefficient for DNA at 260nm wavelength is 20. Based on this extinction coefficient, the absorbance at 260nm is a 1 - cm quartz cuvette of a 50μg/ml solution of double - stranded DNA is equal to 1.

The transition between double stranded and single stranded forms of the moiecules may produce certain changes on the level of absorption, but this deviation can be calibrated with certain formula. This method is convenient and relatively accurate.

Proteins absorb maximally at approximately 280nm due mainly to tryptophan residues. The ratio of A_{260}/A_{280}, therefore, is a measure of the purity of a DNA preparation and should fall between 1. 65 and 1. 85. A lower value suggests protein contamination. If phenol, which has one λ_{max} at 270nm, is contaminating the DNA preparation, then the A_{260} will be abnormally high, leading to an overestimation of the DNA concentration.

Reagents and apparatus

1. DNA sample.

2. Matching pairs of 1 - ml quartz cuvettes.

Note: Glass or plastic cuvettes absorb UV, so quartz cuvettes must be used to measure DNA and RNA concentrations.

3. Eppendorf tubes (Microcentrifuge tubes).

4. Racks for Eppendorf tubes.

5. UV spectrophotometer (UV lamp prewarmed).

Procedures

1. Make 1 : 20 or higher dilution of the stock DNA solution in TE (or ddH_2O). Use TE (or ddH_2O) as blank to calibrate the UV spectrophoto - meter at 260, 280, and 310nm wavelength.

2. Load each dilution of DNA solution into the cuvette and read the optical density (*OD*) at all three wavelengths.

3. Record the *OD* reading and determine the purity and concentration of DNA by calculation.

(1) For ssDNA, the concentration, in microgram per ml, can be derived by the formula:

$$ssDNA = 33(OD_{260} - OD_{310}) \times (\text{dilution factor})$$

(2) For dsDNA, the concentration, in microgram per ml, can be derived by the formula:

$$dsDNA = 50(OD_{260} - OD_{310}) \times (\text{dilution factor})$$

(3) For ssRNA, the concentration, in microgram per ml, can be derived by the formula:

$$ssRNA = 40(OD_{260} - OD_{310}) \times (\text{dilution factor})$$

Notes

1. OD_{310} is the background, this will be high if the salt concentration is high.

2. OD_{260}: OD_{280} for DNA should be around 1. 8. Ratio above or below this usually indicates the presence of RNA or protein contamination in the DNA sample.

3. OD_{260}: OD_{280} for RNA is around 2. 0.

实验三　碱去垢剂法分离质粒 DNA——小量制备法

扫码“学一学”

扫码“看一看”

【导言】

质粒是几乎存在于所有细菌中除染色体之外的环状 DNA 分子。质粒通常携带染色体上不存在的基因，并表现出一些有用的性状，如抗生素抗性、抗细菌等。质粒拥有自己的复制原点，因此可以不依赖于染色体而进行独立复制。一些小的质粒利用宿主细胞的酶进行复制，而较大的质粒则携带编码自身与复制有关的酶的基因。质粒有严紧型（低拷贝数）与松弛型（高拷贝数）之分。不同的质粒可以根据它们所携带的基因以及这些基因所赋予的宿主细胞的性状进行分类，包括抗性质粒（R）、致育因子（F）和 col 以及毒性质粒。

质粒通常用作克隆载体，而碱裂解法是制备质粒 DNA 最为常用的方法之一。本实验可从少量细胞中分离质粒 DNA，称为小量制备法。质粒 pBR329 含有编码三种抗生素的抗性基因，可编码抗氨苄西林（Ap^r）、抗氯霉素（Cm^r）和抗四环素（Tc^r）。一些限制性内切酶位点在 pBR329 中只出现一次。

【目的】

通过本实验掌握碱去垢剂法分离质粒 DNA 的方法。

【原理】

本方法是依据共价闭合环状质粒 DNA 与染色体 DNA 在变性和复性特性之间存在差异进行的。当细胞悬浮于 NaOH 和十二烷基磺酸钠（SDS）溶液中时，在高 pH（碱）的作用下细胞发生裂解，蛋白质和染色体 DNA 发生变性，加入中和溶液（酸性乙酸钾溶液）并离心后，变性蛋白和染色体 DNA 就会与细胞碎片一起沉淀，而质粒 DNA 则留在上清液中。当加入碱溶液时质粒 DNA 也发生变性，但其两条链仍然靠得很近，就像一条链上的两个环链，当加入酸性溶液进行中和时，质粒 DNA 的两条链就分别与其互补链重新退火，进而形成原始状态的质粒。

下面介绍一种简便快速的从少量培养液或大量细胞克隆中分离 DNA 的方法。这种方法提取的质粒 DNA 纯度很高，无须进一步纯化就可以用于限制性酶消化、PCR、序列测定等。

【试剂与器材】

1. 将含有 pBR329 质粒的 HB101：Tn5 菌株在 LB 琼脂平板上划线，37℃培养过夜。
2. 消毒的微量离心管。
3. 冰浴。
4. 溶液Ⅰ（25mM Tris－HCl，pH 8.0，50mM 葡萄糖，10mM EDTA），其中含有

4mg/ml 的溶菌酶，须新鲜配制，冰上保存，临用时加热至室温。

5. 溶液Ⅱ（0.2N NaOH，1.0% SDS，临用前新鲜制备）。

6. 溶液Ⅲ（3M 醋酸钾，3M 醋酸），冰上保存。

7. TE 缓冲液，含有 50μg/ml RNA 酶（80℃热处理 10 分钟，以灭活其中的 DNA 酶）。

8. 异丙醇（室温）。

9. 70% 乙醇。

10. 移液枪。

11. TE 缓冲液饱和的苯酚：氯仿（1：1）溶液。

12. 旋涡混合器。

13. 37℃水浴。

14. 微型离心管。

15. 真空干燥装置。

【实验步骤】

1. 从 LB 平板上挑取一单菌落，接种于 100ml 含相应抗生素的 LB 培养液中，37℃振荡（约 225 转/分钟）培养过夜。

2. 在微型离心管中加入 1.2ml 培养液，12 000×g 离心 30 秒。

3. 移去上清，将沉淀用 100μl 含有 4mg/ml 溶菌酶的溶液Ⅰ重新悬浮（室温下）。在旋涡混合器上将细胞振荡混匀，并确认无粘连的细胞块。

4. 在室温下放置数秒，可见到溶菌酶对细胞的作用：悬浮的细胞变成轻微的乳状，振荡时易附着于试管壁。将试管置于冰浴冷却 1 分钟。

5. 室温下加入 200μl 溶液Ⅱ，轻微振荡试管使其混合，并置试管于冰浴再冷却 1 分钟。

6. 加入 150μl 冰冷的溶液Ⅲ，混合，再冰浴 3 分钟，此时有白色絮状沉淀物形成。

7. 微型离心机 12 000×g 离心 2 分钟。

8. 将上清液移至新的微量离心管中，注意避免带入白色沉淀物。

9. 加入等体积的苯酚：氯仿（1：1）混合物。旋涡混合 1 分钟，室温下 12 000×g 离心 2 分钟。

10. 将上层水相移入新的离心管中，并加入等体积氯仿溶液，旋涡混合振荡 30 秒，12 000×g 离心 2 分钟。

11. 将上层水相移入新的离心管中，加入 2 倍体积的冰冷的无水乙醇，混合后于 −20℃沉淀 5 分钟。

12. 12 000×g 离心 4 分钟，弃去上清，用冰冷的 70% 的乙醇淋洗沉淀，然后对沉淀进行真空干燥。

13. 将沉淀重新悬浮在 50μl 含有 50μg/ml RNA 酶的 TE 缓冲液中。37℃保温 5 分钟。如果不用 RNA 酶进行处理，小于 600bp 的 DNA 片断在琼脂糖凝胶电泳时会被 RNA 所掩盖。

14. 在小量制备的 pBR329/HB101：Tn5 质粒试管标注日期和内容，−20℃保存备用。

Experiment 3　Isolation of plasmid DNA by the alkaline – detergent method——miniprep procedure

Introduction

Plasmids are circular DNA molecules present in addition to the bacterial chromosome in almost all bacteria. Plasmids carry genes not found on the bacterial chromosome which often confer useful properties, such as resistance to antibiotics, to the bacterium. Plasmids have their own origin of replication and so can replicate independently of the chromosome DNA. Some small plasmids use cell's enzymes to replicate. Larger plasmids may carry genes that encode their own replicative enzymes. Plasmids may be stringent(low copy number) or relaxed(high copy number). Many different plasmids can be classified according to the genes they carry and they confer on the host cells, they include resistance(R), fertility(F), col, and virulence plasmids.

Plasmids are frequently used as cloning vectors and alkaline lysis is one of the most widely used methods for the preparation of plasmid DNA. In this exercise, you will isolate plasmid DNA from a small number of cells by this method; the DNA isolated in this manner is often referred to as a miniprep. The plasmid to be isolated, pBR329 contains genes encoding resistance to three antibiotics: ampicillin(Ap^r), chloramphenicol(Cm^r), and tetracycline(Tc^r). Several restriction enzyme cloning sites appear only once in pBR329.

Purpose

By the end of this laboratory exercise, you should be able to isolate plasmid DNA by the alkaline – detergent method.

Principle

This method exploits the difference in denaturation and renaturation characteristics of covalently closed circular plasmid DNA and chromosomal DNA. When cells are suspended in a solution of NaOH and sodium dodectyl sulfate(SDS), the cells are lysed by the high(alkaline) pH, additionally the protein and chromosome DNA will be denatured, and precipitate together with cell debris during the centrifugation after the neutralizing solution(acidic potassium acetate) is added, leaving in the supernatant. With the alkaline added, plasmid DNA is also denatured, but the two strands remain in close proximity, like two links of a chain. Upon neutralization, plasmid DNA strands reanneal to their complementary strands and the plasmid is reformed.

The following procedure is an easy and rapid way for DNA isolation from small scale culture, or many cell colonies at same time. The plasmid DNA achieved by this method is of high purity, therefore can be used for restriction digestion, PCR, sequencing etc. without further purification.

Reagents and apparatus

1. LB agar plate streaked with strain HB101 : Tn5 containing pBR329 and incubated at 37℃

overnight.

2. Sterile microcentrifuge tubes.

3. Ice bucket with ice.

4. Solution Ⅰ(25mM Tris - HCl, pH 8.0, 50mM glucose, 10mM EDTA). Solution Ⅰ containing 4mg/ml lysozyme, freshly prepared (store on ice but warm to room temperature when ready to use).

5. Solution Ⅱ (0.2N NaOH, 1.0% SDS, made fresh just prior to use).

6. Solution Ⅲ (3M potassium acetate, 3M acetic acid) stored on ice.

7. TE buffer containing 50μg/ml RNase A (preheated to 80℃ for 10minutes to inactivate DNases).

8. Isopropanol (room temperature).

9. 70% ethanol.

10. Pipette.

11. Phenol : chloroform (1 : 1) saturated with TE buffer.

12. Vortex mixer.

13. 37℃ water bath.

14. Microcentrifuge.

15. Vacuum drying apparatus.

Procedures

1. Inoculate a single colony from an LB plate in 100ml of LB medium containing the appropriate antibiotic. Incubate overnight at 37℃ with shaking (approximately 225rpm).

2. Place 1.2ml of the overnight culture into a microcentrifuge tube and centrifuge at 12 000 × g for 30 seconds.

3. Remove the supernatant and resuspend the pellet in 100μl of Solution Ⅰ containing 4mg/ml lysozyme (at room temperature). Vortex cells into suspension, making sure there are no clumps of cells.

4. After several seconds at room temperature, lysozyme action on the cells will be visible: the suspended cells turn slightly milky and tend to coat the sides of the tubes when shaken. Place the tubes into an ice bucket and chill 1minute.

5. Add 200μl of Solution Ⅱ at room temperature and mix by inversion. Return tube to the ice bucket for another minute.

6. Add 150μl of ice - cold Solution Ⅲ, mix, and return to the ice bucket for another 3minutes. A fluffy white precipitate will form.

7. Centrifuge at 12 000 × g for 2minutes in a microcentrifuge.

8. Transfer the suspematant to a fresh tube, avoiding the white precipitate.

9. Add an equal volume of phenol : chloroform (1 : 1). Vortex for 1minute and centrifuge at 12 000 × g for 2minutes at room temperature.

10. Transfer the upper, aqueous phase to a fresh tube and add an equal volume of chloroform. Vortex for 30 seconds and centrifuge at 12 000 × g for 2minutes.

11. Transfer the upper, aqueous phase to a fresh tube and add 2 volumes of ice - cold 100% ethanol. Mix and allow toprecipitate 5minutes at -20℃.

12. Centrifuge at 12 000 × g for 4minutes. Pour off the supernatant, rinse the pellet with ice - cold 70% ethanol and dry the pellet under vacuum.

13. Resuspend the pellet in 50μl of TE buffer containing 50μg/ml of RNase A. Incubate at 37℃ for 5minutes. In the absence of RNase treatment, DNA fragment smaller than about 600bp are obscured by RNA when separated by agarose gel electrophoresis.

14. Label the tube containing the miniprep of pBR329/HB101 : Tn5 with date and contents, and store at -20℃.

扫码“学一学”

扫码“看一看”

实验四　琼脂糖凝胶电泳

【导言】

电泳常用于分离，有时也用于纯化分子大小、电荷性状或分子构象不同的生物大分子，尤其是蛋白质和核酸，因此，电泳已成为生物化学和分子生物学中应用最广泛的技术之一。带电分子置于电场中，根据它们所带电荷的不同，可向正极（阳极）或负极（阴极）移动。蛋白质与核酸不同，蛋白质分子可以带正电也可以带负电，而核酸只能带负电，这是由其磷酸盐分子骨架所决定的，且只能向阳极泳动。生物大分子是在某种基质或“凝胶”中迁移的，根据制备凝胶所用材料的不同，凝胶电泳又分为两类：琼脂糖和聚丙烯酰胺凝胶电泳。琼脂糖凝胶电泳是分离和纯化 DNA 片断应用最普遍的技术。

【目的】

通过本实验掌握琼脂糖凝胶电泳的操作方法。

【原理】

琼脂糖是一种海藻多糖，琼脂糖胶分离范围很大，但其分辨率却相对较低。通过改变琼脂糖凝胶的浓度，应用标准的电泳技术可以分离 200 ~ 50 000bp 大小的 DNA 片断。琼脂糖凝胶浓度越大，凝胶就越硬。较高浓度的琼脂糖胶有利于较小的 DNA 分离，而较低浓度的琼脂糖胶则可以分离较大的 DNA 片断。琼脂糖胶浓度一般在 0.5% ~2.0% 之间。通过观察示踪染料的迁移距离可以判断 DNA 的迁移距离。溴酚蓝和二甲苯青染料在琼脂糖凝胶中的迁移速率大致分别与 300 和 4000bp 大小的双链 DNA 片断相同。迁移足够距离后，就可以通过溴化乙锭染色来观察 DNA 片断。溴化乙锭是一种荧光染料，它嵌插在 DNA 和 RNA 碱基之间。在做胶时将其混入其中在电泳时进行染色，也可以待电泳完成后将凝胶浸泡在稀释的溴化乙锭溶液中进行染色。必须将凝胶置于紫外分析检测器中才可以对凝胶中的 DNA 或 RNA 进行观察。

【试剂与器材】

1. 琼脂糖。

2. TAE（Tris－乙酸，EDTA）[50 倍的浓缩储备液：Tris，242g；冰醋酸 57.1ml；0.5M EDTA（pH 8.0），100ml；加入 600ml 蒸馏水，剧烈振摇，再加入蒸馏水至 1000ml]。

3. 溴化乙锭储备液（10mg/ml）：在 100ml 水中加入 1g 溴化乙锭，用磁力搅拌器搅拌数小时，转移到黑色瓶中，4℃保存（溴化乙锭是强诱变剂，在称取时务必戴手套和面罩。一旦接触要立即用大量的水冲洗）。

4. TBE [5 倍储备液（1L）：54gTris 碱，27.5g 硼酸，20ml 0.5M EDTA，pH 8.0]。

5. 10×加样缓冲液：0.25% 溴酚蓝，0.25% 二甲苯青 FF，15% 聚蔗糖水溶液。

6. 实验器材：广口烧杯，量筒，搅拌棒，微波炉，托盘天平，凝胶板，梳子，电泳槽，电源。

【实验步骤】

（一）凝胶的制备

1. 搭好凝胶电泳槽。梳齿应该保持笔直，在梳齿底部与胶槽之间应保持数毫米的间隙。

2. 称取适量琼脂糖，加入 Tris－乙酸缓冲液中，用微波炉加热至溶解（如称取 1.0g 琼脂糖，溶于 100ml Tris－乙酸缓冲液中，制成 1% 的琼脂糖）。

3. 将琼脂糖溶液缓缓倒在插有齿梳的凝胶板上，避免产生气泡。如果有气泡，用巴斯德吸管去除。目前，许多类型的凝胶板都可以购买，按照说明书操作即可。传统的经济简便的方法是用胶带将合适大小的塑料板或玻璃板的边缘封住，形成一个胶模，以此作为凝胶板。可用手指在凝胶板边缘来回触压数次，确证密封完全，以防琼脂糖泄漏。

4. 根据凝胶的大小，让其凝固 20～60 分钟。

5. 在琼脂糖凝固过程中，准备好 DNA 样品，并向其中加入含有染料（如溴酚蓝）的加样缓冲液。

6. 琼脂糖凝胶凝固之后，移走齿梳，将凝胶板移至电泳槽上，加样孔在靠近阴极的一端（黑色端）。向槽中加入适量的 TAE 缓冲液，通常没过胶面 1cm。再加入溴化乙锭，混合完全（最终浓度达到 1μg/ml）。另一种方法是将溴化乙锭加入凝胶中，方法为：用微波炉将琼脂糖煮沸，然后冷却至 50℃，加入溴化乙锭。

（二）加样、电泳

1. 用移液枪缓慢将 DNA 样品垂直加入加样孔直至其开口下方。

2. 加完所有样品后，将电泳槽与电源正确连接（黑色对阴极，红色对阳极）。打开电源之前调好电压和时间。

3. 观察铂丝是否已连接到加样孔附近的黑色端以确保导线连接正确。如果导线连接正确，会有气泡缓慢上升。

4. 电泳过程需要一定的时间，时间长短取决于凝胶的长度及使用的电压大小。凝胶越长、电压越低，则所需时间就越长。但是在分离较大的 DNA 片断时，使用高电压效果较差。

（三）凝胶（图像）解释

未切割质粒 DNA 在其泳道上可能会出现几个条带，这是由于质粒 DNA 在琼脂糖凝胶中的迁移距离是由其分子构象及碱基对大小所决定的。质粒 DNA 可以下列 3 种主要构象中的任何一种形式存在。

1. 超螺旋 尽管质粒通常以开环的形式进行描述，然而在细菌细胞内 DNA 链却是盘绕在组蛋白样蛋白质周围形成一种致密结构。这就是超螺旋结构，由于其结构致密，它在凝胶中的泳动速度最快。

2. 切口 在质粒 DNA 复制过程中，拓扑异构酶Ⅰ会在 DNA 双螺旋中的一条链中引入一个切口，解开质粒的超螺旋。在质粒分离过程中由于物理剪切和酶的切割作用也会在超螺旋质粒中引入切口从而产生松散的开环结构。这种形式的质粒迁移速率最慢，其“松散”的分子形式阻碍了它在琼脂糖凝胶中的运动。

3. 线性 当 DNA 损伤在 DNA 双链相对应的两条链上同时产生切口时，就会出现线性质粒 DNA。这种 DNA 的泳动速率介于超螺旋与切口质粒 DNA 之间。质粒制备过程中出现线性 DNA 说明存在核酸酶污染或实验操作有问题。

（四）凝胶摄影

电泳之后，在紫外光下可观察到 DNA 片断在琼脂糖凝胶中得到分离，同时可对凝胶进行拍照。市面上可购买到各种照相机，在说明书的指导下很容易操作。大体上，要拍好一张照片，照相机的光圈孔径和感光率是关键因素。例如，为了增加曝光度，使影像密度更高，需要加长曝光时间，这可通过减慢快门速度或增加感光率得以实现。与之相反，若要降低曝光度，使影像更模糊，则要缩短曝光时间或减小透镜孔径。

【注意事项】

1. TAE 和 TBE 均为常用的缓冲液。TBE 比 TAE 有相对高的缓冲能力。

2. 加样染料溴酚蓝可与长度约为 0.5kb 的 DNA 一起迁移，可用于指示迁移率最高的片断。

3. DNA 的迁移速率取决于以下因素。

（1）DNA 的分子大小 分子质量越小，迁移越快。

（2）琼脂糖浓度 浓度越低，迁移越快。

（3）DNA 构象 环状或带切口环状 DNA 通常比线状 DNA 迁移快。

（4）两个电极之间单位厘米的电压 电压越高，迁移越快。

4. 如果 DNA 条带不够窄且不够均匀，可能是以下原因所引起。

（1）DNA 过载。

（2）电压过高。

（3）加样孔破损。

（4）凝胶中有气泡。

Experiment 4 Agarose gel electrophoresis

Introduction

Electrophoresis is a technique used to separate and sometimes purify macromolecules, especially proteins and nucleic acids, which differ in size, charge or conformation. As such, it is one of the most widely used techniques in biochemistry and moleoular biology. When charged molecules are placed in an electric field, they migrate toward either the positive (anode) or negative (cathode) pole according to their charge. In contrast to proteins, which can have either a net positive or net negative charge, nucleic acids have a consistent negative charge imparted by their phosphate backbone, and migrate toward the anode. Macromolecules are electrophoresed within a matrix or 'gel'. According to the difference of materials used to make the gel, gel electrophoresis may be divided into two classes: agarose and polyacrylamide electrophoresis. Agarose gel electrophoresis is now the most popular technique used to separate and purify DNA fragments.

Purpose

By the end of this laboratory exercise, you should be able to describe how to run an agarose gel.

Principle

Agarose is a polysaccharide extracted from seaweed. Agarose gels have a large range of separation, but relatively low resolving power. By varying the concentration of agarose, fragments of DNA from about 200 – 50 000bp can be separated using standard electrophoretic techniques. The higher the agarose concentration, the "stiffer" the gel. Higher concentrations of agarose facilite separation of small DNAs, while low agarose concentrations allow resolution of larger DNAs. It is typically used at concentrations of 0.5% – 2.0%. The distance DNA has migrated in the gel can be judged by visually monitoring migration of the tracking dyes. Bromophenol blue and xylene cyanol dyes migrate through agarose gels at roughly the same rate as double – stranded DNA fragments of 300 and 4000bp, respectively. When adequate migration has occured, DNA fragments can be visualized by staining with ethidium bromide. This fluorescent dye intercalates between bases of DNA and RNA. It is often incorporated into the gel so that staining occurs during electrophoresis, but the gel can also be stained after electrophoresis by soaking in a dilute solution of ethidium bromide. Tovisualize DNA or RNA, the gel is placed on a ultraviolet transilluminator.

Reagents and apparatus

1. Agarose.

2. TAE(Tris – Acetate, EDTA)[concentrated stocksolution 50X: Tris base, 242g; glacial acetic acid, 57.1ml; 0.5M EDTA(pH 8.0), 100ml; add to 600ml dH_2O, stir vigorously, bring to 1000ml with dH_2O].

3. Ethidium bromide stock(10mg/ml): add 1g of ethidium bromide to 100ml of H_2O, stir with a magnetic stirrer for several hours and trasfer to a dark bottle and store at 4℃ (Ethidium bromide is a powerful mutagen. Wear gloves and a mask when weighting it out. In case of contact, immediately flush with copious amounts of water).

4. TBE[5 × stock solution(1L): 54g Tris base, 27.5g boric acid, 20ml 0.5M EDTA, pH 8.0].

5. 10 × loading buffer: 0.25% bromophenol blue, 0.25% xylene cyanol FF, 15% Ficoll in water.

6. Equipment: beaker, graduated cylinder, stir bar, microwave, Pan balace, gel bed, comb, electrophoresis tank, and power supply.

Procedures

1. Preparation of the gel

(1) Set up the gel apparatus as demonstrated. The comb should be straight, and there should be a few millimeters of clearance between the bottom of the comb and the bottom of the gel tray.

(2) Weigh proper amount of agarose (e.g. 1.0g agarose in 100ml TAE buffer makes 1% agarose) and dissolve in TAE buffer by heating it in a microwave oven.

(3) Pour the agarose slowly (aviod bubbles) onto a gel bed with the comb inserted. Use the

plastic transfer pipette to remove the bubble if there is any. Currently, many varieties of gel beds are available commercially and those can be used according to the instruction. However, the "traditional", as well as economical and simple gel bed can be made by sealing a proper sized plastic or a glass with the masking tape. Make sure the seal is tight to prevent agarose from leaking, and this can be achieved by running your finger along the edge of the gel bed several times.

(4) Let gel polymerize for about 20 – 60minutes depending upon the size of the gel.

(5) While the agarose gel is polymerizing, prepare the DNA sample and mix proper amount of DNA with the loading buffer containing the dye such as bromophenol blue.

(6) Remove the comb after the agarose gel has polymerized and place the gel bed onto the electrophoresis tank with the wells near the cathode (black terminal). Fill the tank with proper amount of TAE buffer. Usually the buffer is about 1cm above the gel. Add ethidium bromide (final concentration is 1μg/ml) to the TAE buffer and mix it well. Alternatively, ethidium bromide can be added to the agarose gel. This can be done by boiling the agarose in the microwave and cooling to about 50℃ and then adding ethidium bromide.

2. Loading DNA samples and gel running

(1) Hold the pipette perpendicular to the well and add DNA sample slowly with the pipette tip just beneath the opening of the well.

(2) After all the samples have been loaded to the wells, connect the gel tank with the power supply properly (black to cathode [–] and red to anode [+]). Set the voltage and time before turning on the power supply.

(3) Make sure the leads have been properly connected by watching the platinum wire at the black terminal near the wells. Bubbles should be slowly rising if the leads have a good connection.

(4) Allow electrophoresis to progress for appropriate time. The timing of electrophoresis depends on the length of the gel and the amount of voltage applied. The longer the gel and the lower the voltage, the longer time is needed. However, high voltages are significantly less effective at resolving large DNA fragments.

3. Gel interpretation

The uncut DNA lane may have several bands in it. This occurs because the migration of plasmid DNA in an agarose gel depends on its molecular conformation as well as its size in base pairs. Plasmid DNA can exist in any one of three major conformations:

(1) Supercoiled: Although a plasmid is usually pictured as an open circle, within a bacterial cell the DNA strand is coiled around histone – like proteins to form a compact structure. This is called supercoiling and this form of the plasmid will move the fastest through the gel due to its compact structure.

(2) Nicked: During plasmid DNA replication, topoisomerase I introduces a nick into one strand of the DNA helix and uncoils the plasmid. Physical shearing and enzymatic cleavage during plasmid isolation may also introduce nicks into the supercoiled plasmid to produce a relaxed open circular structure. This form is the slowest migrating form of plasmid. Its "floppy" molecular shape impedes movement through the agarose gel.

(3) Linear: Linear plasmid DNA occurs when damage results in strand nicks directly opposite

each other on the DNA helix. This DNA moves at a rate intermediate between supercoiled and nicked plasmid DNA. The presence of linear DNA in a plasmid preparation is a sign of either nuclease contamination or sloppy lab procedure.

4. Gel photography

DNA fragments separated on the agarose gel after electrophoresis can be visualized and photographed under the UV light. Many kinds of cameras are available commercially and they can be easily manipulated following the instructions. In principle, the aperture and the speed are the two key factors for taking a good picture. For example, to increase exposure, i. e. to make image more intense, a longer exposure time is required and this can be achieved by slowing the shutter speed, or increasing the aperture. In contrast, to decrease exposure, i. e. to make pictures darker, a shorter exposure time or smaller lens opening is required.

Notes

1. Both TAE and TBE are common buffer used. TBE has relative higher buffering capacity than TAE.

2. The loading dye bromophenol blue migrates with DNA about 0. 5kb and provides an index of the mobility of the fastest fragments.

3. The migration of the DNA depends on the following factors.

(1) Molecular size of DNA: the smaller the DNA, the faster the migration.

(2) Agarose concentration: the lower the concentration, the faster the migration.

(3) Conformation of the DNA: circular or nicked DNA often migrate faster: than that of linear DNA.

(4) Voltage per cm distance between electrodes: the higher the voltage, the faster the migration.

4. If the DNA bands are not sharp and uniform, it may be due to the following reasons.

(1) Overloaded DNA.

(2) Voltage too high.

(3) Torn well.

(4) Bubble in gel.

扫码“学一学”

扫码“看一看”

实验五　聚合酶链反应（PCR）

【导言】

聚合酶链反应（PCR）技术是 Kary Mullis 和他的同事于 20 世纪 80 年代发明的，该技术相对简单，可以快速可靠地将 DNA 或 cDNA 模板扩增数千倍乃至上百万倍。PCR 利用一对寡核苷酸引物扩增一段 DNA 序列，其中每条引物分别与靶序列的两端互补。这对引物通过三步循环反应在热稳定 DNA 聚合酶的作用下向各自对应的方向延伸，每步循环反应包括 3 个步骤：模板变性、引物退火和引物延伸。

【目的】

通过本实验了解 PCR 的工作原理和操作步骤。

【原理】

如图 5－1 所示，PCR 可以利用 DNA 聚合酶对一定区域内的 DNA 进行扩增。可以在所需要扩增的 DNA 片段（X）两侧设计两个 DNA 短片段（引物），这两个引物可以分别与 X 区域两侧杂交，并引发（引导）DNA 合成。X 的两条链的拷贝以及原来的 DNA 链则可以充当下一轮扩增反应的模板。这样，所选定区域 DNA 的数量就可以随着每一轮循环而倍增，达到起始数量的数百万倍，直至可以通过凝胶电泳进行肉眼检测。最初，研究人员在每轮循环都必须添加新鲜的 DNA 聚合酶，因为标准的酶不能耐受每轮复制前的 DNA 变性高温（90℃以上）。然而，现在已经有专门的可以耐受高温的热稳定聚合酶。其中，*Taq* 聚合酶来源于一种存在于温泉中的细菌 *Thermos aquaticus*，因此其聚合酶可以耐热。只需将 *Taq* 聚合酶与引物以及模板在试管中进行混合、封口，然后置于热循环仪中，设计程序使热循环仪在 3 个不同温度间反复循环：首先，高温（约 95℃）使 DNA 双链解离；其次，在相对较低的温度（约 55℃）下使引物与模板 DNA 链退火；最后，在适中的温度（约 72℃）下使 DNA 合成。每轮循环只需几分钟，通常不到 20 个循环就可以扩增出足够量的 DNA。

典型的扩增反应包括靶 DNA 样品，热稳定 DNA 聚合酶，两条寡核苷酸引物，dNTPS（各种三磷酸脱氧核苷酸），反应缓冲液，Mg^{2+} 及其他添加物。反应液组分必须进行混合，然后置于可以自动控制不同温度与时间的热循环仪中进行反应。

PCR 技术目前已成为分子生物学中克隆与基因分析的基本工具。

【试剂】

1. 灭菌水。
2. 10 × 扩增缓冲液，含 15mM $MgCl_2$。
3. 10mM dNTPs。
4. 50μM 寡核苷酸引物 1。

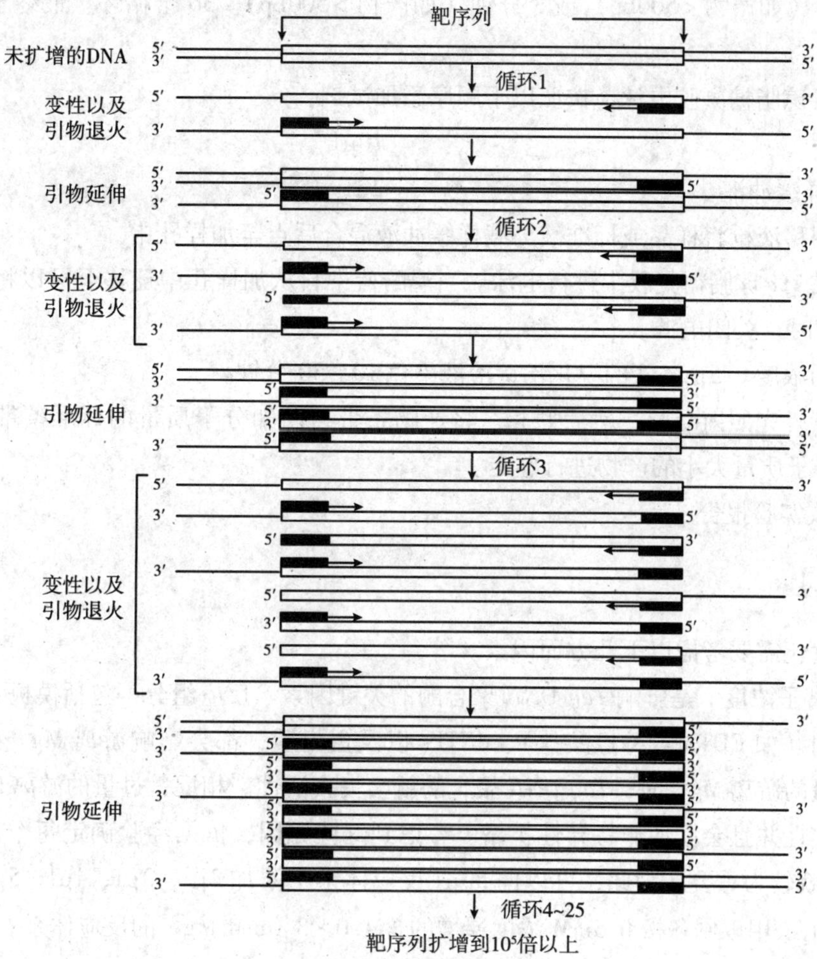

图 5-1　PCR 反应过程示意

5. 50μM 寡核苷酸引物 2。
6. 5unit/μl *Taq* DNA 聚合酶。
7. 模板 DNA（1μg 基因组 DNA，0.1～1ng 质粒 DNA）。

【实验步骤】

1. 在 PCR 管中将下列各组分混合：

10×PCR 缓冲液	10μl
引物 1	1μl
引物 2	1μl
dNTPs	2μl
模板 DNA 和水	85.5μl
Taq DNA polymerase	0.5μl

2. 设立对照反应，其中不含模板 DNA，而加 10μl 水。
3. 向每个反应管中添加 70～100μl 矿物油（或 2 滴硅油）。
4. 将反应管置于热循环仪中预热到 94℃。
5. 按以下程序运行：94℃ 1 分钟→55℃（或适合于特定引物对的退火温度）1 分钟→

72℃ 1 分钟（如产物<500bp）或 3 分钟（如产物>500bp）。30 个循环。最终于 72℃ 延伸 7 分钟。

6. 进行琼脂糖凝胶电泳，按照以下顺序添加试剂：

产物	10μl
6×载样缓冲液	2μl

（1）用移液枪将样品或标准液与载样缓冲液混合后点至加样孔中。

（2）在 1% 琼脂糖凝胶中进行电泳。当溴酚蓝染料从加样孔中迁移至可以将 DNA 片段分开的距离时，关闭电源。

（3）用浓度 0.5μg/ml 的溴化乙锭溶液染色 10～30 分钟。

（4）紫外光照射凝胶，然后照相。将产物条带与已知分子质量的标准条带进行比较，可对合适分子质量大小的产物进行鉴定。

7. 完成实验报告。

【注释】

PCR 优化需要考虑以下几方面因素。

1. 镁离子浓度 是影响 *Taq* DNA 聚合酶的关键因素。反应组分，包括模板 DNA、样品中的螯合剂（如 EDTA 或柠檬酸盐）、dNTPs 以及蛋白质，都会影响游离 Mg^{2+} 的浓度。当缺少足够量的游离 Mg^{2+} 时，*Taq* DNA 聚合酶就会失去活性。相反，过量的游离 Mg^{2+} 则会降低酶的忠实性并且会增加非特异性扩增。考虑到这些原因，依据经验确定每个反应的最佳 $MgCl_2$ 浓度就较为重要。为此，可以向 50μl 反应体系中添加 2μl、3μl、4μl、5μl 或 6μl 的 25mM $MgCl_2$，用以制备按 0.5mM 浓度递增的含 1.0～3.0mM Mg^{2+} 的反应体系。

2. 酶浓度 在 50μl 扩增反应体系中推荐使用 1.25 单位的 *Taq* DNA 聚合酶。大多数情况下，这将导致酶过量，而较多的酶也不会显著增加产物量。增加酶量以及过度增加延伸时间会增加出现假带的可能性，而这与 *Taq* DNA 聚合酶的 5′→3′核酸外切酶活性有关，最终导致琼脂糖凝胶的 smear 现象。

3. 引物设计 PCR 引物长度一般为 15～30 碱基，设计的两引物之间为 DNA 目的区域。引物设计时其 G+C 含量应在 40%～60%，注意避免在引物内部出现二级结构。两引物间的 3′末端彼此间不应互补，以免在 PCR 反应中产生引物二聚体。引物 3′末端应避免 3 个 G 或 C 核苷酸连续排列。较为理想的是两条引物在相同的温度下退火，而退火温度则取决于其中熔点温度（T_m）较低的那条。可用以下公式粗略计算 T_m 值：

$$T_m = 2℃ \times (\text{A 和 T 残基数}) + 4℃ \times (\text{G 和 C 残基数})$$

4. 循环次数 通常在 25～35 之间。较多的循环虽然意味着较高的产量，然而随着循环数的增加，产生各种错误产物（如引导错误的产物）的概率也增加。一般不超过 40 个循环。

5. 核酸交叉污染 尽量减少样品间的交叉污染并防止将一个实验中的 RNA 和 DNA 带到下一个实验。应将扩增前和扩增后的工作区以及移液枪分开使用，应戴手套并经常更换。

Experiment 5　Polymerase chain reaction (PCR)

Introduction

Polymerase chain reaction(PCR) was invented by Kary Mullis and his colleagues in the 1980s, it is a relatively simple technique by which a DNA or cDNA template is amplified many thousand or million fold quickly and reliably. PCR is used to amplify a sequence of DNA, using a pair of oligonucleotide primers each complementary to one end of the DNA target sequence. These are extended towards each other by a thermostable DNA polymerase in a reaction cycle of three steps: denaturation of template, annealing of primers and extension of the primers.

Purpose

By the end of this laboratory exercise, you should be able to explain the principle of PCR and describe how it runs.

Principle

As Fig. 5 – 1 explains, this technique uses enzyme DNA polymerase to make a copy of a defined region of DNA. We can select the part(X) of the DNA we want to amplify by putting in short pieces of DNA(primers) that hybridize to DNA sequences on either side of X and cause initiation (priming) of DNA synthesis through X. The copies of both strands of X, as well as the original DNA strands, then serve as templates for the next round of amplification. In this way, the amount of the selected DNA region double over and over with each cycle, up to millions of times the starting amount, until there is enough to see by gel electrophoresis. Originally, worker had to add fresh DNA polymerase at every round because standard enzymes do not stand up to the high temperatures (over 90℃) needed to separate the strands of DNA before each round of replication. However, we now have special heat stable polymerase that can take the heat. One of these, *Taq* polymerase, comes from *Thermus aquaticus*, a bacterium that lives in hot springs and therefore has heat – stable enzymes. All we have to do is mix the *Taq* polymerase with the primers and template DNA in a test tube, seal the tube, than place it in a thermal cycler. We program the thermal cycler to cycle over and over again among three different temperatures: first a high temperature(about 95℃) to separate the DNA strands; then a relatively low temperature(about 55℃) to allow the primers to anneal to the template DNA strands; then a medium temperature(about 72℃) to allow DNA synthesis. Each cycle takes as little as a few minutes, and it usually takes fewer than 20 cycles to produce as much amplified DNA as we need.

A typical amplification reaction includes the sample of target DNA, a thermostable DNA polymerase, two oligonucleotide primers, dNTPS(deoxynucleotide triphosphates), reaction buffer, Mg^{2+} and optional additives. The components of the reaction are mixed and the reaction is placed in a thermal cycle, which is an automated instrument that takes the reaction through a series of different

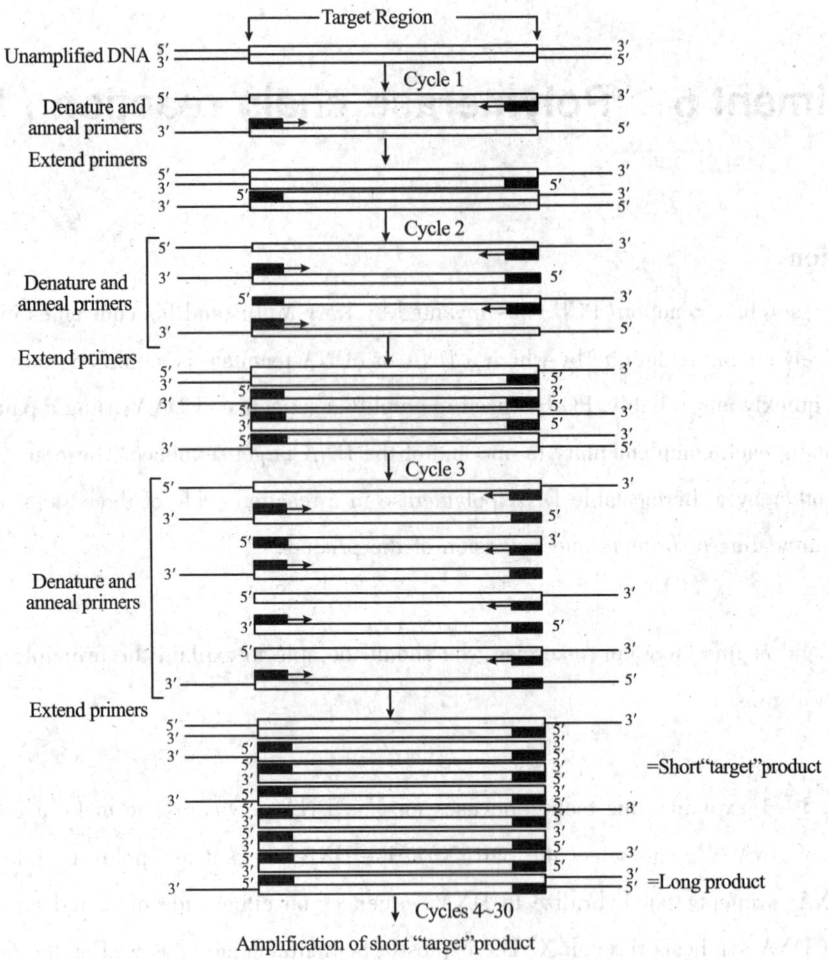

Fig. 5-1 Schematic Diagram of the PCR Process

temperatures for varying amounts of time.

PCR technique has become an essential tool in molecular biology as an aid to cloning and gene analysis.

Reagents and apparatus

1. Sterile water.
2. 10 × amplification buffer with 15mM $MgCl_2$.
3. 10mM dNTPs.
4. 50μM oligonucleotide primer 1.
5. 50μM oligonucleotide primer 2.
6. 5unit/μl *Taq* DNA polymerase.
7. template DNA(1μg genomic DNA,0.1-1ng plasmid DNA).

Procedures

1. Combine the following for each reaction in a PCR tube:

10 × PCR buffer	10μl
Primer 1	1μl
Primer 2	1μl

dNTPs	2μl
Template DNA and water	85. 5μl
Taq DNA polymerase	0. 5μl

2. Prepare a control reaction with no template DNA and an additionl 10μl of sterile water.

3. Add 70 – 100μl mineral oil(or 2 drops of silicone oil) to each reaction tube.

4. Place tubes in a thermal cycler preheated to 94℃.

5. Run the following program: 94℃ 1min→55℃ 1min or annealing temperature appropriate for particular primer pair→72℃ 1min(if product is <500bp) or 3min(if product is >500bp) for 30 cycles. Program a final extension at 72℃ for 7min.

6. Electrophorese 10μl on an agarose gel.

Product	10μl
6 × loading buffer	2μl

(1) Load samples or marker mixed with loading buffer into wells with a pipettor.

(2) Electrophorese on an 1% agarose gel. Turn off the power supply when bromphenol blue dye from the loading buffer has migrated a distance judged sufficient for separation of DNA fragments.

(3) Stain with 0. 5μg/ml ethidium bromide solution for 10 to 30minutes.

(4) Illuminate the gel with UV light and then photograph. By comparing product bands with bands from the known molecular – weight markers, you should be able to identify any product fragments which are of the appropriate molecular weight.

7. Finish your report exercises.

Notes

Considerations for PCR optimization include following aspects.

1. Magnesium concentration

Magnesium concentration is a crucial factor affecting the performance of *Taq* DNA polymerase. Reaction components, including template DNA, chelating agents present in the sample(e. g. EDTA or citrate), dNTPs and proteins, all affect the amount of free magnesium. In the absence of adequate free magnesium, *Taq* DNA polymerase is inactive. Conversely, excess free magnesium reduces enzyme fidelity and may increase the level of nonspecific amplification. For these reasons, it is important to empirically determine the optimal $MgCl_2$ concentration for each reaction. To do so, prepare a reaction series containing 1. 0 – 3. 0mM Mg^{2+} in 0. 5mM increments by adding 2μl, 3μl, 4μl, 5μl or 6μl of the 25mM $MgCl_2$ stock to 50μl reactions.

2. Enzymes concentration

It is recommended that using 1. 25units of *Taq* DNA Polymerase in a 50μl amplification reaction. For most applications, this results in enzyme excess and the inclusion of more enzyme does not significantly increase product yield. It should also be noted that increased amounts of enzyme and excessively long extension times increase the likelihood of generating artifacts associated with the intrinsic 5′→3′ exonuclease activity associated with *Taq* DNA Polymerase, resulting in smearing in agarose gels.

3. Primer design

PCR primers generally range in length from 15 – 30 bases and are designed to flank the region of interest. Primers should contain 40% – 60% G + C and care should be taken to avoid sequences which would produce internal secondary structure. The 3′ – ends of the primers should not be complementary to avoid the production of primer – dimers in the PCR reaction. Avoid three G or C nucleotides in a row near the 3′ – end of the primer. Ideally, both primers should anneal at the same temperature. The annealing temperature is dependent upon the primer with the lowest melting temperature (T_m). A simple formula for estimating the T_m:

$$T_m = 2℃ \times (\text{number of A and T residues}) + 4℃ \times (\text{number of G and C residues})$$

4. Number of cycles

The number of cycles is usually between 25 and 35. More cycles mean a greater yield of product. However, with increasing number of cycles the greater the probability of generating various artifacts (e. g. mispriming products). It is unusual to find procedures which have more than 40 cycles.

5. Nucleic acid cross – contamination

It is important to take great care to minimize the potential for cross – contamination between samples and to prevent carryover of RNA and DNA from one experiment to the next. Use separate work areas and pipettors for pre – and post – amplification steps. Wear gloves and change them often.

实验六　限制性内切酶与 DNA 的消化

扫码“学一学”

扫码“看一看”

【导言】

Ⅱ型限制性内切酶是一类可以识别并结合到双链 DNA 的特异核酸序列上同时对其进行水解切割的酶。至今已有几千种限制性内切酶被分离出来，表现出几百种不同的序列（底物）特异性，通常根据其来源细菌的名称对它们加以命名。它们是现代分子生物学中进行 DNA 序列操作与鉴别的主要工具之一。

【目的】

通过本实验掌握限制性内切酶的工作原理与方法。

【原理】

限制性酶在脱氧核糖和磷酸基团间对 DNA 进行水解，结果分别在两条双链的末端留下 5′端磷酸基团和 3′端羟基。有少数限制性酶可以切割单链 DNA，尽管其切割效率很低。

核酸限制性内切酶一个特性是其所识别的序列是回文结构，即两条链在 5′→3′方向上分别具有相同的序列。典型的限制性酶切位点（Ⅱ型）是一个由 4bp、5bp、6bp、7bp 或 8bp 组成的具有一个旋转对称轴的精确回文结构（如 *Eco*R Ⅰ的识别序列为 GAATTC）。一些限制性酶在 DNA 双链识别中心周围做交错对称切割，另一些酶在识别位点中心做对称切割。前一种情况产生 DNA 黏性末端，后一种情况产生 DNA 钝端。

在进行限制性酶消化时，将双链 DNA 分子与适量的限制性酶在供应商推荐的相应的缓冲液中进行混合保温，并在该酶的最适温度下进行反应。典型的消化反应包括每毫克起始 DNA 一个单位的酶，一个酶单位通常定义（依据供应商）为一小时内在合适的温度下，尤其是 37℃下，用于完全消化一毫克双链 DNA 的酶的数量。为了保证消化完全，通常需要保温 1～3 小时。

【试剂与器材】

1. 未线性化的质粒 pBR329。
2. 限制性酶 *Eco*R Ⅰ。
3. 限制性酶缓冲液。
4. 10×*Eco*R Ⅰ缓冲液。

（1）100mM Tris－HCl，25℃时 pH 7.5。

（2）50mM NaCl。

（3）10mM 氯化镁。

（4）0.025% Triton X－100。

5. 实验三制备的质粒样品。

6. DNA 大小标准。

7. 琼脂糖凝胶电泳装置。

【实验步骤】

1. -20℃冰箱中取出实验三制备的质粒样品，让其升至室温。

2. 限制性酶消化是在特定的缓冲液中进行的。缓冲液以 10×浓缩贮存液供应，使用时必须稀释成 1×反应液。反应可以在 20μl 总体积中进行。

3. 在标有“E”的试管中加入 10μl 质粒 DNA，7μl ddH_2O，2μl *Eco*R Ⅰ 10×缓冲液，1μl *Eco*R Ⅰ酶液。

4. 加入所有组分后，盖上盖子轻弹混匀并在微量离心机上离心 2~3 秒，将试剂甩至管底。

5. 将试管置于 37℃水浴保温 2 小时。

6. 与此同时，按实验四方法铺琼脂糖胶。

7. 将反应样品与 5μl 载样染料混匀准备点样。

8. 按以下顺序点样，并在笔记本上记录加样顺序：

泳道	样品
A	DNA 大小标准
B	*Eco*R Ⅰ限制性消化的样品
C	未切割的质粒

9. 以 10V/cm 的电压进行电泳直至深蓝染料到达凝胶末端（约 1 小时）。

10. 电泳跑完后请助理教师给凝胶拍照。如果凝胶未出结果，或许还要进行重新消化或重新电泳。

11. 将限制性酶切割的质粒泳道与未切割质粒的泳道进行比较。它们应有所不同，否则，消化反应很可能就有问题。假如不能断定，可以向实验老师请教。

12. 未切割质粒 DNA 的泳道也许有多个条带，这与质粒 DNA 在琼脂糖凝胶中的迁移率和其分子构象以及大小有关。质粒 DNA 可以呈现三种主要构型中的一种，见实验四。

13. 测定酶切质粒泳道的质粒片段的大小。只有线型（消化的）质粒 DNA 分子在凝胶中的泳动速率与其大小呈反比，因此不可根据迁移距离来确定未进行切割的质粒的大小。

Experiment 6 Restriction endonuclease and digestion of DNA

Introduction

Type Ⅱ restriction endonucleases are enzymes that recognize, bind to and hydrolyze specific nucleic acid sequences in double-stranded DNA. Several thousand different restriction endonucleases have been isolated, which collectively exhibit a few hundred different sequence (substrate) specificities. They are commonly named after the bacterium from which it was isolated. They are one of the primary tools in modern molecular biology for the manipulation and identification of DNA sequences.

Purpose

Upon completion of this exercise you should be able to describe how restriction endonuclease works.

Principle

Restriction enzymes hydrolyze the backbone of DNA between deoxyribose and phosphate groups. This leaves a phosphate group on the 5′ ends and a hydroxyl on the 3′ ends of both strands. A few restriction enzymes will cleave single stranded DNA, although usually at low efficiency.

One unique feature of restriction enzymes is that the nucleotide sequences they recognize are palindromic, i. e. they are the same sequences in the 5′→3′ direction of both strands. The typical restriction enzyme site(Type Ⅱ) is an exact palindrome of 4bp, 5bp, 6bp, 7bp or 8bp with an axis of rotational symmetry(e. g. the *Eco*R Ⅰ recognition site is GAATTC). Some restriction endonucleases make staggered symmetrical cuts away from the center of their recognition site within the DNA duplex, some make symmetrical cuts in the middle of their recognition site. Enzymes that make staggered cuts leave the resultant DNA with cohesive or sticky ends. Enzymes that cleave the DNA at the center of the recognition sequence leave blunt – ended fragments of DNA.

Restriction enzyme digestions are performed by incubating double – stranded DNA molecules with an appropriate amount of restriction enzyme, in its respective buffer as recommended by the supplier, and at the optimal temperature for that specific enzyme. Typical digestions included a unit of enzyme per microgram of starting DNA, and one enzyme unit usually(depending on the supplier) is defined as the amount of enzyme needed to completely digest one microgram of double – stranded DNA in one hour at the appropriate temperature, typically 37℃. To insure complete digestion the reactions usually are incubated for 1 – 3hours.

Reagents and apparatus

1. Unlinearized plasmid pBR329.
2. Restriction enzymes *Eco*R Ⅰ.
3. Buffers for restriction enzymes.
4. 10 × Buffer for *Eco*R Ⅰ.

(1) 100mM Tris – HCl, pH 7.5 at 25℃.

(2) 50mM NaCl.

(3) 10mM Magnesium Chloride.

(4) 0.025% Triton X – 100.

5. Samples from Expt. 3.
6. DNA Size Markers.
7. Apparatus for Agarose Gel Electrophoresis.

Procedures

1. Remove the DNA samples from Expt. 3 from – 20℃ refrigerator. Allow these samples to reach room temperature.

2. Restriction enzyme digests are performed in a special buffer. The buffer is supplied in a 10 × concentrated stock, you will have to dilute it to 1 × in your reactions. You will be performing your digestions in a total volume of 20μl.

3. To the tube labeled "E", add 10μl of plasmid DNA, 7μl of ddH_2O, 2μl of 10 × buffer for *Eco*R Ⅰ, and 1μl of *Eco*R Ⅰ.

4. After adding all ingredients, cap the tubes, flick them to mix the contents and pool the reagents in the bottom by spinning the tubes in the microcentrifuge for 2 to 3 seconds.

5. Incubate the tubes in a 37℃ water bath for 2 hours.

6. Meanwhile, pour an agarose gel as described in the Expt. 4.

7. To prepare your samples for electrophoresis, mix your samples with 5μl of loading dye.

8. Load the samples onto the gel in the following order and record the order of your samples in your notebook:

Lane	Sample
A	DNA Size Markers
B	Restriction Digest – *Eco*R Ⅰ
C	Uncut Plasmid

9. Electrophorese your gel at 10V/cm until the dark blue dye has reached the end of the gel (about 1hour).

10. After the gel has finished running, has the T. A. take a picture of it for you. If your gel didn't turnout, you may have to redigest or rerun your samples.

11. Compare the lanes in which you digested the plasmid with enzymes to the lane containing the uncut DNA. They should look different. If they do not, your digestion probably did not work. Ask if you are unsure.

12. The uncut plasmid DNA lane may have several bands in it. This occurs because the migration of plasmid DNA in an agarose gel depends on its molecular conformation as well as its size in base pairs. Plasmid DNA can exist in any one of three major conformations.

13. Determine the sizes of the fragments in the lanes in which the plasmid was digested with enzyme. Only linear (digested) DNA moves through the gel at a rate that is inversely proportional to its size in base pairs. Therefore you cannot determine the size of a plasmid based on the migration of uncut plasmid DNA.

实验七　DNA 连接与 TA 克隆

扫码“学一学”

扫码“看一看”

【导言】

DNA 连接是线性 DNA 片断以共价键连在一起的过程，它在重组 DNA 技术中至关重要。TA 克隆则是一种快速的、一步就可以将 PCR 产物直接插入质粒载体的方法。

【目的】

学习掌握 DNA 连接和 TA 克隆的技术与方法。

【原理】

连接包括在一个核苷酸的 3′端羟基与另外一个核苷酸的 5′端磷酸基团之间形成磷酸二酯键的过程。一般采用 T_4DNA 连接酶连接 DNA 片断，它来源于 T_4噬菌体。T_4DNA 连接酶可以连接具有单链突出的黏性末端片断，也可以连接具有平齐末端的片断，只是此时通常需要较高浓度的连接酶。除了水之外连接反应通常还需要 3 种组分：两个或两个以上的具有匹配的黏性末端或平齐末端的 DNA 片断；含有 ATP 的缓冲液；T_4DNA 连接酶。

TA 克隆利用 PCR 反应中的热稳定聚合酶（即 *Taq* 聚合酶），可以不依赖模板向 PCR 产物 3′末端加上一个腺苷酸（A）。这些末端突出的 3′A 即可用于将 PCR 产物直接插入插入位点具有一个 3′T 突出的线性 T 载体分子。这一方法省略了用 Klenow 酶或 T_4多聚激酶对 PCR 产物进行修饰以使其形成钝端或添加插入用的接头等一些费时的酶法修饰过程。

【试剂与器材】

1. 具有 T 突出的 T－载体（如 Promega 公司的 pGEM © －T 载体）。
2. *Taq* 聚合酶扩增得到的 PCR 产物。
3. T_4 DNA 连接酶。

【实验步骤】

1. 在一个微量离心管中建立以下连接混合物：

25μg/ml 扩增的靶 DNA	1μl
具有 T 尾巴的质粒	20ng
10×连接缓冲液	1μl
T_4DNA 连接酶（3Weiss 单位/微升）	1μl
加 H_2O	至终体积 10μl

如果需要可添加 ATP 至 1mM 终浓度。载体与扩增的 DNA 片段的摩尔比以 1∶5 为宜。建立一个对照反应，其中除了不含有扩增的靶 DNA 外，含有以上所有列出的反应物。

2. 将连接混合物于 16℃连接 12 小时。

3. 分别用 10μl H_2O 将 5μl 连接混合物进行稀释，然后采用实验八的方法转化合适的 *E. coli* 感受态细胞，并将转化的细胞培养物涂布在含有 IPTG 和 X－gal 以及相应的抗生素的平板上。

4. 计算由连接物转化得到的菌落数。挑取一定量由连接反应转化得到的含有靶基因的白色菌落。确认靶基因片段的存在，分离质粒 DNA 并用多克隆位点中的基因片段两端的限制性酶切位点进行切割。

蓝白菌落数之比一般为 1∶5 至 2∶1 不等。

5. 通过琼脂糖凝胶电泳对限制性片段进行分离，并以合适大小的 DNA 标准作为参照，以确定克隆片段的大小。

6. 通过 DNA 序列测定、限制性酶谱分析或 Southern 杂交对克隆片段进行鉴定。

【注释】

1. 除了具有 3′端到 5′端核酸外切酶活性的聚合酶（如 Vent 和 pfu 酶）扩增得到的产物由于克隆效率极低之外，任何条件下的 PCR 产物都可以用作 TA 克隆。除非产物含有较多的杂质，否则并不需要进行清洁或纯化。

2. T_4DNA 连接酶 10×缓冲液含有 ATP，如果温度波动会发生降解，因此使用时应分装成只供一次使用的小包装，以免反复冻融。

3. 如果融化的 10×缓冲液出现沉淀，就必须对其进行旋涡混合以使沉淀重新溶解（旋涡混合前可在手指间滚动试管，有助于使溶液升温）。

4. 为使只具有单个碱基突出的黏性末端退火充分，连接反应应在低温下进行。如果连接温度高于 15℃会使重组子数目锐减。

Experiment 7　Ligation of DNA and TA cloning

Introduction

DNA Ligation is to join linear DNA fragments together with covalent bonds which accomplishes the "stitching together" of two disparate fragments of DNA. TA cloning is a quick, one－step method for the direct insertion of a PCR product into a plasmid vector.

Purpose

The purpose of this exercise is to learn about the method for ligation of DNA and TA cloning.

Principle

More specifically, DNA ligation involves creating a phosphodiester bond between the 3′ hydroxyl of one nucleotide and the 5′ phosphate of another. The enzyme used to ligate DNA fragments is T_4 DNA ligase, which originates from the T_4 bacteriophage. This enzyme ligate DNA fragments having overhanging, cohesive ends. T_4 DNA ligase also ligate fragments with blunt ends, although higher concentrations of the enzyme are usually recommended for this purpose. A ligation reaction requires three ingredients in addition to water: Two or more fragments of DNA that have either compatible cohesive ("sticky") or blunt ends; a buffer which contains ATP; T_4 DNA ligase.

The TA Cloning takes advantage of the template independent addition of a single adenosine (A) to the 3′-ends of PCR products by activity of thermostable polymerases (i. e. *Taq* polymerase) used in PCR. These 3′A-overhangs are used to insert the PCR product directly into the T vector which is a linear molecule and contains single 3′ Toverhangs at its insertion site. This procedure eliminates time consuming enzymatic modifications of the PCR product such as treating with Klenow or T_4 polymerase to create blunt ends, adding adapter for insertion etc..

Reagents and apparatus

1. T-vector with T overhang (such as pGEM© -T vector, Promega Corporation).
2. Taq polymerase amplified PCR product.
3. T_4 DNA ligase.

Procedures

1. In a microfuge tube, set up the following ligation mixture:

25μg/ml amplified target DNA	1μl
T-tailed plasmid	20ng
10 × ligation buffer	1μl
T_4 DNA ligase (3 Weiss units/μl)	1μl
H_2O	to a final volume of 10μl

If necessary, add ATP to a final concentration of 1mM. A 1∶5 molar ratio of vector: amplified DNA fragment is recommended. Set up a control reaction that contains all the reagents listed above except the amplified target DNA.

2. Incubate the ligation mixture for 12hours at 16℃.

3. Dilute 5μl of each of the two ligation mixtures with 10μl of H_2O and transform a suitable strain of competent *E. coli* as described in Expt. 8. Plate the transformed cultures on media containing IPTG, X-gal and the appropriate antibiotic.

4. Calculate the number of colonies obtained from each of the ligation mixtures. Pick a number of white colonies obtained by transformation with the ligation reaction containing the target DNA. Confirm the presence of the amplified fragment by isolating the plasmid DNAs and digesting them with restriction enzymes whose sites flank the insert in the multiple cloning site.

The ratio of blue and white colonies varies between 1∶5 and 2∶1.

5. Fractionate the restricted DNA by electrophoresis through an agarose gel using appropriate DNA size markers.

Measure the size of the cloned fragments.

6. Confirm the identity of the cloned fragments by DNA sequencing, restriction mapping, or Southern hybridization.

Notes

1. PCR products under any conditions may be used for TA cloning but polymerases with extensive 3′ to 5′ exonuclease activity (such as Vent and pfu) should not be used due to extremely low

cloning efficiencies. There is no need to clean or purify PCR product unless the products contain multiple species.

2. T_4 DNA Ligase 10 × Buffer contains ATP which degrades during temperature fluctuations. Avoid multiple freeze thaws by making single - use aliquots of the buffer.

3. If a precipitate forms, vortex the thawed T_4 DNA Ligase 10 × Buffer until the precipitate is back in solution (It may help to warm the solution by rolling the tube between your fingers before vortexing).

4. Low temperature ligations are necessary for annealing of single base overhangs. Ligation temperatures higher than 15℃ may significantly reduce the number of recombinants.

实验八　感受态大肠埃希菌的制备与转化（氯化钙法）

扫码“学一学”

扫码“看一看”

【导言】

细菌转化是一种重要的分子生物学现象。它使得体外构建的重组 DNA 分子以及天然 DNA 分子可以进行增殖与基因表达。转化还具有重要的历史意义，因为它导致了 20 世纪 40 年代后期 Avery 的发现——DNA 是遗传物质。

【原理】

细菌细胞由于吸入纯化的 DNA 而发生转化，转化也是将重组 DNA 分子导入大肠埃希菌最常用的手段。DNA 在通常情况下不能穿过大多数细菌的细胞膜，为了将 DNA 透过细胞膜导入细胞，细胞必须处于感受态。就大肠埃希菌而言，感受态细胞可以在加入 DNA 之前用氯化钙处理诱导而成。Ca^{2+} 使细胞膜失去稳定性，并形成磷酸钙 - DNA 复合物，黏附在细胞表面。当细胞进行短暂 42℃ 热休克处理时，便会吸入 DNA。冰冷氯化钙处理可诱导细菌产生感受态，而热休克则可以打开细胞膜使 DNA 进入细胞。本实验所介绍的方法简单快速，每毫克超螺旋质粒 DNA 可以产生多达 10^7 个转化子。此后该技术又做了许多改进，目的都是提高转化效率。转化效率是指每 1 微克 DNA 转化所得到的转化细胞数。实际上所使用的 DNA 数量要少得多（通常为 5 ~ 100ng），因为过多的 DNA 会抑制转化过程。

【目的】

学习掌握氯化钙制备大肠埃希菌感受态细胞以及质粒转化的方法。

【试剂与器材】

1. *E. coli* $DH_5\alpha$。
2. 质粒 DNA。
3. 37℃ 培养箱。
4. 37℃ 空气振荡器。
5. 微型离心机。
6. 离心管（灭菌）。
7. 基本培养基。
8. LB 培养基：10g 蛋白胨，5g 酵母提取物，5gNaCl，加水至 1L，高压灭菌。
9. LB 抗性平板：每升 LB 培养基加入 15g 琼脂，高压蒸汽灭菌。冷却至 55℃ 时加入抗生素，倒入培养皿。

【实验步骤】

（一）感受态细胞制备

1. 大肠埃希菌在 LB 平板上划线涂布，倒置平板于 37℃ 培养 1 ~2 天。

2. 从 LB 平板上挑取一单菌落接种到 2.5ml LB 培养液中，37℃振荡培养过夜（约 250rpm）。

3. 第二天，向装有 100ml LB 培养基的 500ml 锥形瓶中转接 2ml 过夜细胞培养物。37℃振荡培养至 OD_{600} 值达 0.5～0.6（需 2～3 小时）。

4. 4℃ 4500×g 离心 5 分钟收集细胞。用 0.6 倍体积于原初培养液 100mM 的冰冷 $CaCl_2$ 重悬细胞，冰浴 30 分钟。

5. 4℃ 4500×g 离心 5 分钟收集细胞。

6. 用原初培养液的 1/25 倍体积的 100mM 的冰冷 $CaCl_2$ 重悬细胞。

7. 将试管置于冰上以待加入 DNA。

注：感受态细胞对外界处理与高温特别敏感，因此操作要温和。

（二）感受态细胞转化

1. 为了对 $CaCl_2$ 处理过的细胞直接进行转化，用冷的微量移液枪枪头移取 200μl 感受态细胞至无菌的冷的 17mm×100mm 聚丙烯试管中。向试管中加入 DNA（不超过 50ng，体积不超过 10μl）。轻度旋涡混合后冰上放置 30 分钟。

2. 将试管转至已预热到 42℃循环水浴中的试管架上，精确放置 90 秒。不要摇动试管。

3. 将试管迅速转移至冰浴，使细胞冷却 1～2 分钟。

4. 向试管中加入 800μl LB 培养基。37℃水浴培养 45 分钟以恢复和表达由质粒编码的抗生素抗性标记。

5. 将适量体积（每块 90mm 平板不超过 200μl）的转化细胞转移至含有适当抗生素的 LB 琼脂培养基上。

6. 室温放置平板直至液体被吸收。

7. 倒置平板于 37℃培养 12～16 小时后应出现转化菌落。

【注释】

1. 获得高感受态细胞的关键是在全过程中使细胞保持冰冻。

2. 为获得高感受态，应使用对数生长期的细胞。因此 OD_{600} 值不应高于 0.6。

3. 感受态大肠埃希菌细胞很脆弱，处理时必须小心。

Experiment 8 Preparation and transformation of competent *E. coli* using calcium chloride

Introduction

Bacterial transformation is a phenomenon of importance in molecular biology. It allows for the propagation and genetic expression of recombinant DNA molecules that have been constructed *in vitro* as well as natural DNA molecules. Transformation is also of historical importance since it led to the discovery by Avery, in the late 1940's, that DNA was the genetic material.

Purpose

Learn how to prepare competent *E. coli* Using calcium chloride and transform it with the plasmid.

Principle

Transformation results from the uptake of purified DNA by bacterial cells and is the most frequently used procedure for introducing recombinant DNA molecules into*E. coli*. However, DNA is normally unable to cross the membranes of most bacteria. In order to make bacteria bring DNA into the cell through their membranes, they must be made competent. Competence can be artifically induced in *E. coli* cells by treating them with calcium chloride prior to adding DNA. The Ca^{2+} destabilizes the cell membrane, and a calcium phosphate – DNA complex is formed, which adheres to the cell surface. The DNA is taken up during a heat – shock step when the cells are exposed briefly to a temperature of 42℃. It is believed that the ice – cold calcium chloride treatment induces a "competence" state in the bacteria and the heat shock opens the cell membrane to let DNA enter the cell. This is a simple and quick method which yields up to 10^7 transformed colonies per microgram of supercoiled plasmid DNA. Many variations of this technique have since been developed, all directed toward optimizing the transformation efficiency. Transformation efficiency is the amount of cells transformed per 1 microgram of DNA. In practice, much smaller amounts of DNA are used (5 to 100ng) since too much DNA inhibits the transformation process.

Reagents and apparatus

1. *E. coli* $DH_5\alpha$.
2. Plasmid DNA.
3. 37℃ incubator.
4. 37℃ air – shaker.
5. Microfuge.
6. Centrifuge tubes (sterile).
7. Minimal media.
8. LB medium: 10g tryptone, 5g yeast extract, 5g NaCl. Add H_2O to 1liter, autoclave.
9. LB plates with antibiotics: Add 15g agar per liter LB medium, autoclave. After cooling down to 55℃, add antibiotics. Pour onto Petri dishes.

Procedures

1. Competent cell preparation

(1) Streak *E. coli* cells onto minimal media plate, invert the plate and grow at 37℃ for 1 – 2days.

(2) Inoculate a single colony from an LB plate in 2.5ml of LB medium. Incubate overnight at 37℃ with shaking (approximately 250rpm).

(3) On the following day, Transfer 2ml of the entire overnight culture to inoculate 100ml of LB medium in 500ml flask, grow the cells at 37℃ with shaking to OD_{600} between 0.4 and 0.6 (takes 2 – 3hours).

(4) Pellet the cells by centrifugation at 4500 × g for 5minutes at 4℃. Gently resuspend the cell pellets in 0. 6 volume (based on the original culture volume) of ice – cold 100mM $CaCl_2$. Incubate the resuspended cells on ice for 30minutes.

(5) Pellet the cells by centrifugation at 4500 × g for 5minutes at 4℃.

(6) Gently resuspend the cell pellets in 1/25 of the original culture volume of ice – cold 100mM $CaCl_2$.

(7) Place the tubes on ice before you are ready to add DNA.

Note: Treat the competent cells gently as they are highly sensitive to handing and elevated temperature.

2. Transformation of competent cells

(1) To transform the $CaCl_2$ – treated cells directly, transfer 200μl of each suspension of competent cells to a sterile, chilled 17mm × 100mm polypropylene tube using a chilled micropipette tip. Add DNA (no more than 50ng in a volume of 10μl or less) to each tube. Mix the contents of the tubes by swirling gently. Store the tubes on ice for 30minutes.

(2) Transfer the tubes to a rack placed in a preheated 42℃ circulating water bath. Store the tubes in the rack for exactly 90 seconds. Do not shake the tubes.

(3) Rapidly transfer the tubes to an ice bath. Allow the cells to chill for 1 – 2minutes.

(4) Add 800μl of LB medium to each tube. Incubate the cultures for 45minutes in a water bath set at 37℃ to allow the bacteria to recover and to express the antibiotic resistance marker encoded by the plasmid.

(5) Transfer the appropriate volume (up to 200μl per 90 – mm plate) of transformed competent cells onto agar LB medium containing the appropriate antibiotic.

(6) Store the plates at room temperature until the liquid has been absorbed.

(7) Invert the plates and incubate at 37℃. Transformed colonies should appear in 12 – 16hours.

Notes

1. It is critical to keep cells ice cold throughout this procedure to obtain high competency.

2. For high competency, cells in log phase should be used. Therefore, the OD_{600} should not higher than 0. 6.

3. Competent *E. coli* cells are fragile and must be treated carefully.

实验九　利用 α 互补筛选细菌克隆

扫码“学一学”

扫码“看一看”

【导言】

任何转化实验都只有一小部分的细菌细胞被转化。质粒上的抗生素标记可快速筛选转化子。只有转化成功的细胞，由于携带抗生素抗性标记，可以在含有抗生素的培养基上生存和生长。许多广泛使用的克隆质粒载体还存在第二层次的筛选途径——α 互补，即蓝－白选择，利用这种方法可以确定具有 DNA 插入片断的质粒。

【目的】

通过本实验掌握利用 α 互补筛选细菌克隆的原理与方法。

【原理】

大肠埃希菌 β－半乳糖苷酶的两个无活性片断结合可形成具有功能的酶，产生 α 互补。许多质粒载体携带 β－半乳糖苷酶前 146 个氨基酸残基编码信息的 DNA 短片断。这类载体一般与表达 β－半乳糖苷酶羧基末端部分的宿主菌配套使用。尽管宿主菌与质粒编码的 β－半乳糖苷酶片断各自都没有酶活性，但它们结合成一体，形成具有酶活性的蛋白。由于 α 互补产生 Lac^+ 的细菌很容易识别，因为它们在显色底物 5－溴－4－氯－3－吲哚－β－半乳糖苷（X－gal）存在时会形成蓝色。然而，在质粒的多克隆位点插入外源 DNA 片断后，几乎不可避免地导致无 α 互补能力的氨基端片断产生。因此，载有重组质粒的细菌就不能水解 X－gal 并形成白色菌落。这一简单的实验方法很大程度上简化了质粒载体重组子的鉴定过程。

【试剂与器材】

1. *E. coli* $DH_5\alpha$－宿主细胞。
2. *E. coli* $DH_5\alpha$－插入片断与载体（pUC19）连接混合物转化的细胞。
3. IPTG 溶液（20%，*W/V*）。
4. X－gal 溶液（2%，*W/V*）。
5. 含有合适抗生素的丰富肉汤琼脂平板。
6. 丰富肉汤顶层琼脂。

【实验步骤】

（一）方法一

1. 将融化的顶层琼脂分装于 17mm×100mm 的试管中。将试管置于 45℃加热保温备用。

2. 将第一支试管取出，快速加入 0. 1ml 的细菌悬液。如使用 90mm 的平板，其中的活细菌数小于 3000；如使用 150mm 的平板，其中的活细菌数小于 10 000。堵住试管口并颠倒数次使细菌在顶层琼脂中分布均匀。

3. 打开试管，按表 9－1 加入适量 X－gal 和 IPTG。堵住试管口并缓和颠倒试管数次以将其中的内含物混合均匀。

表 9－1 顶层琼脂组成

平板大小	琼脂溶液量	X－gal	IPTG
90mm	3ml	40μl	7μl
150mm	7ml	100μl	20μl

4. 快速将融化的顶层琼脂倒在已经变硬的含有适当抗生素的琼脂平板的中心，旋转使其分散。

5. 重复第 2～4 步直至铺完所有样品。

6. 在室温下使软琼脂变硬，擦去平板盖上的冷凝水，于 37℃倒置培养 12～16 小时。

7. 从培养箱中取出平板于 4℃储存数小时，使蓝颜色变深。

8. 对含有重组质粒的菌落进行鉴别。携有野生型质粒的菌落，具有 β－半乳糖苷酶活性，其中心为淡蓝色；外周为深蓝色；携有重组质粒的菌落，无 β－半乳糖苷酶活性，呈乳白色或蛋壳蓝色，其中心为淡蓝色，有时中心会出现一个微弱的蓝点。

9. 挑选含有重组质粒的克隆并进行培养。

（二）方法二

直接在琼脂平板上用 X－gal 和 IPTG 进行筛选。该方法不需要制备顶层琼脂，它是将 X－gal 的浓缩溶液涂布在预先制备好的琼脂平板上，而不是将卤代半乳糖苷掺入整个琼脂培养基。涂布 X－gal 溶液时必须小心。由于整个平板的 X－gal 浓度存在差异，处于平板中心的菌落也许蓝色会加深。

1. 吸取 40μl 2% 的 X－gal 溶液和 7μl 20% 的 IPTG 溶液，点于 90mm 含有合适抗生素的 LB 或 YT 平板的中央。对于直径 15cm 的平板，则需要转移 100μl 的 X－gal 溶液和 20μl 的 IPTG 溶液。

2. 使用无菌涂布器（或弯曲的巴斯德吸管，其吸头已在火焰上封口）将溶液涂至整个平板表面，将平板置于 37℃培养直至表面所有液体消失。由于二甲基甲酰胺挥发性较低，如果平板是现做的，这一过程一般需要 3～4 小时。

3. 可用接种环或牙签将待检测的细菌接种到平板上，也可将 100μl 的细菌悬液（50000 个细胞/毫升）直接涂布于 90mm 琼脂平板表面，或 200μl 涂布于 15cm 琼脂平板表面。

4. 待接种物吸收之后，将平板于 37℃倒置培养 12～19 小时。

5. 将平板从培养箱取出并于 4℃贮存数小时，在此期间蓝色会充分显现。

按照上述两种方法处理后，对含有重组质粒的菌落进行鉴定。

Experiment 9 Screening bacterial colonies using α－complementation

Introduction

During any transformation, only a small percentage of the bacterial cells present will be transformed. The presence of an antibiotic marker on the plasmid allows for rapid screening of successful transformants. Only those cells successfully transformed, and thus carrying the antibiotic resistance gene, will be able to survive and grow on antibiotic containing medium. Many of the widely used plasmid cloning vectors carry a second level of selection, α – complementation: a blue – white selection that can be used to determine which plasmids carry an inserted fragment of DNA and which do not.

Purpose

Upon completion of this exercise you should be able to describe how to screen bacterial colonies usingα – complementation.

Principle

α – complementation occurs when two inactive fragments of *E. coli* β – galactosidase associate to form a functional enzyme. Many plasmid vectors carry a short segment of DNA containing the coding information for the first 146 amino acids of β – galactosidase. Vectors of this type are used in host cells that express the carboxy – terminal portion of the enzyme. Although neither the host nor the plasmid – encoded fragments of β – galactosidase are themselves active, they can associate to form an enzymatically active protein. Lac^+ bacteria that result from α – complementation are easily recognized because they form blue colonies in the presence of the chromogenic substrate 5 – bromo – 4 – chloro – 3 – indoyl – β – D – galactoside (X – gal). However, insertion of a fragment of foreign DNA into the polycloning site of the plasmid almost invariably results in production of an aminoterminal fragment that is no longer capable of α – complementation. Bacteria carrying recombinant plasmids therefore cannot degrade X – gal and will form white colonies. The development of this simple blue white color test has greatly simplified the identification of recombinants constructed in plasmid vectors.

Reagents and apparatus

1. *E. coli* $DH_5\alpha$ – host cells.
2. *E. coli* $DH_5\alpha$ – transformed with ligation mixture of insert to vector arms (pUC19).
3. IPTG solution (20% *W/V*).
4. X – gal solution (2% *W/V*).
5. Rich broth agar plates containing the appropriate antibiotic.
6. Rich broth top agar.

Procedures

1. Procedure A

(1) Dispense aliquots of molten top agar into 17mm × 100mm tubes. Place the tubes in a 45℃ heating block until they are needed.

(2) Remove the first tube from the heating block. Working quickly, add 0.1ml of bacterial suspension containing < 3000 viable bacteria for a 90mm plate and < 10 000 for a 150mm plate. Close the top of the tube and invert it several times to disperse the bacteria through the molten agar.

(3) Open the tube and add the appropriate amounts of X – gal and IPTG (if required) as shown

in table 9 – 1. Close the top of the tube and gently invert it several times to mix the contents.

Table 9 – 1 Components for Top Agar

Size of Plate Molten Top Agar	Amount of Reagent	X – gal	IPTG
90mm	3ml	40μl	7μl
150mm	7ml	100μl	20μl

(4) Quickly pour the molten top agar into the center of a hardened agar plate containing the appropriate antibiotic and distribute the solution by swirling.

(5) Repeat Steps 2 – 4 until all of the samples have been plated.

(6) Allow the soft agar to harden at room temperature, wipe any condensation from the lid of the plates, and then incubate the plates in an inverted position for 12 – 16 hours at 37℃.

(7) Remove the plates from the incubator and store them for several hours at 4℃, to allow the blue color to develop.

(8) Identify colonies carrying recombinant plasmids. Colonies that carry wild – type plasmids contain active β – galactosidase. These colonies are pale blue in the center and dense blue at their periphery. Colonies that carry recombinant plasmids do not contain active β – galactosidase. These colonies are creamy – white or eggshell blue, sometimes with a faint blue spot in the center.

(9) Select and culture colonies carrying recombinant plasmids.

2. Procedure B

Direct application of X – gal and IPTG to agar plates. An alternative to preparing top agar can be achieved by spreading a concentrated solution of X – gal on the surface of a premade agar plate, rather than incorporating the halogenated galactoside throughout the entire volume of the agar medium. Take care when spreading the X – gal solution. Colonies in the center of the plate may be a deeper blue due to variations in the concentration of X – gal across the plate.

(1) Pipette 40μl of 2% X – gal solution and 7μl of 20% IPTG solution onto the center of a premade 90mm plate (e. g. LB or YT) containing the appropriate antibiotic. For a 15cm diameter agar plates, transfer 100μl of X – gal and 20μl of IPTG to the center of the plate.

(2) Use a sterile spreader (or a bent Pasteur pipette whose tip has been sealed in a flame) to spread the solutions over the entire surface of the plate. Incubate the plate at 37℃ until all of the fluid has disappeared. Because of the low volatility of dimethyl formamide, this procedure can take up to 3 – 4hours if the plate is freshly made.

(3) Inoculate the plate with the bacteria to be tested by streaking with a bacterial loop, by arranging clones with toothpicks, or by spreading up to 100μl of a bacteria suspension (50 000cells/ml) on the surface of a 90mm agar plate or 200μl on a 15cm plate.

(4) After the inoculum has been absorbed, incubate the plate in an inverted position for 12 – 19hours at 37℃.

(5) Remove the plate from the incubator and store it at 4℃ for several hours, during which the blue color develops to its full extent.

(6) Identify colonies carrying recombinant plasmids.

实验十　Southern 印迹

扫码“学一学”

扫码“看一看”

【导言】

Southern 印迹是第一个以核酸杂交为基础的分析技术，它是由牛津大学的 Edward Southern 发明的。这一技术主要用于检测 DNA 分子及分析基因的结构。

【目的】

1. 检测复杂混合物中特定基因片段的有无。
2. 测定位于特定基因中的 DNA 限制片段长度。

【原理】

Southern 印迹的第一步是用限制性内切酶对 DNA 进行消化，然后再用琼脂糖凝胶电泳将已消化的 DNA 进行分离。电泳后对凝胶进行照相以记录 DNA 分子质量标准品的迁移位置，然后将凝胶浸泡于碱性溶液中进行变性以便进行杂交。第二步是将变性的 DNA 片段从凝胶转移到尼龙膜上（有时用硝酸纤维素膜），这样就可用探针进行分析。把凝胶置于缓冲液浸透了的多孔物质的表面，再将膜置于凝胶上面，再在最上面覆盖干燥的印迹用纸。缓冲液在毛细管效应的作用下通过凝胶转移到最上层的纸上，变性的 DNA 也随着缓冲液向上移动并停留于膜上。这些 DNA 通过烘烤或紫外交联则可永久地结合于膜上。通过印迹，膜上 DNA 的位置（相对于其他片断）就与其原来在凝胶中的位置相同。第三步是将膜与已标记的探针进行温育，即杂交。这一过程是在有利于探针与结合在膜上的 DNA 片断之间形成杂交分子的温度与缓冲液中进行的。杂交后的膜用缓冲液进行洗涤以去除非特异性结合的探针，这样就只有与靶序列结合的探针仍存在于膜上。如果探针用放射性同位素标记，则尼龙膜可以通过放射自显影进行分析。冲洗 X 线片就会显现出一条或多条黑色条带，这些条带与结合了探针的 DNA 片段的位置一致，这样，就可以依据杂交片断的数目以及大小对特定基因的结构进行分析鉴定。

【试剂与器材】

1. 核酸及寡聚核苷酸。
2. 基因组 DNA。
3. 沙门菌 DNA（约 10mg/ml）。
4. 放射性标记 DNA 探针或放射性标记 RNA 探针（比放射强度 10^9cpm/μg）。
5. 限制性内切酶。
6. 1×TAE 缓冲液。
7. 3M NaCl。
8. 10N NaOH。

9. 1M Tris－HCl，pH 8.0。

10. 0.7% 1×TAE 琼脂糖。

11. 10% SDS。

12. 20×SSC（20L）：将 3504g NaCl 和 1760g 柠檬酸盐溶解于 10L ddH_2O 中（使用 20L 的酸坛）。用 ddH_2O 调整体积到 20L，滴加浓盐酸调节 pH 7.4。

【实验步骤】

（一）印迹

1. 用一种或多种限制性内切酶消化适量的基因组 DNA。

2. 如果需要，消化结束后用乙醇沉淀法浓缩 DNA 片段。将 DNA 溶解于 25μl TE 中（pH 8.0）。

3. 用荧光法、溴化乙锭或 SYBR Gold spot 测试法测定已消化 DNA 的浓度。各取适量消化物置于微量管中。加入 0.15 倍体积的 6×蔗糖凝胶加样缓冲液，通过琼脂糖凝胶电泳将 DNA 片段进行分离（对于大多数基因组 DNAs 可以使用 0.7% 的凝胶及 0.5×TBE 或 1×TAE 缓冲系统）。维持低压（约 <1V/cm）以使 DNA 缓慢迁移（如果消化的 DNAs 曾贮藏于 4℃，在加样前应先加热到 56℃并维持 2～3 分钟。加热可以打开有可能出现的黏性末端之间的碱基配对）。

4. 电泳结束后，用溴化乙锭或 SYBR Gold 染色并照相。在凝胶旁放置透明尺，以便直接从相片中读出 DNA 条带的迁移距离。

5. DNA 经电泳分离后，将凝胶转移到玻璃培养皿。用刀片切除空白区，包括加样孔上方的凝胶。注意保留足够的凝胶上的孔，以便 DNA 转移后条带的位置可以在膜上标记出来。在凝胶的左下角切下，以便在后续的操作中起定位作用。切除含有标准分子质量对照的泳道。

6. 将凝胶在新制备的含 0.6M NaCl 和 0.2N NaOH 的溶液（200ml 3M NaCl＋20ml 10N NaOH，ddH_2O 定容至 1000ml）中缓慢振荡 30 分钟以使其中的 DNA 变性。保证凝胶浸于变性液中。如果凝胶表面干燥，DNA 就无法完全变性，会导致转移效果差。可以用洁净的玻璃试管压在凝胶上面保证浸没状态。用 ddH_2O 对凝胶进行简单漂洗。

7. 在含 1.5M NaCl，0.5M Tris－HCl，pH 8.0 的溶液（1L 含 500ml 3M NaCl＋500ml 1M Tris－HCl）中轻微振荡，以中和凝胶。经常检查以保证凝胶浸没。

8. 如果使用已准备好的 Southern 印迹装置（图 10－1），只需去除顶层的印迹支持物，然后加上新近在 10×SSC 中浸泡过的印迹支持物。如果使用新的转移装置，需将印迹支持物依次迅速浸没于 10×SSC 中并将它们转移到新的印迹盘中。印迹支持物叠放达到 1.27cm（0.5 英寸）厚度即可。叠放时，用洁净的试管在表面轻轻滚动以去除层与层之间的气泡。

9. 倒入 10×SSC 的缓冲液，液面高度达到支持物顶部，但不要没过顶端。

10. 按印迹支持物的大小切取 3MM Whatman 滤纸（或使用预先切好的 3MM Whatman 滤纸），分别用 10×SSC 润湿，铺于印迹支持物上。依照上述方法通过滚动试管去除气泡。

11. 用洁净剪刀（接触膜时需要戴手套并且操作时要在洁净的 3MM Whatman 滤纸上进行）切取稍大于凝胶的 Zetabind 膜（或其他作为固体支撑物的尼龙膜）。在膜的上部用记号笔标记各自的序号、酶及印迹号，并切除膜的左上角以便定位。

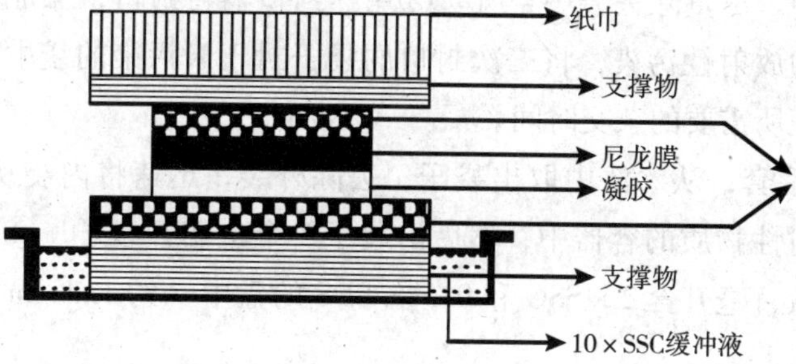

图 10-1 Southern 印迹装置

12. 在 ddH_2O 中短暂润湿尼龙膜，并将其转移到盛有 10×SSC 的盘中，浸泡 20 分钟。

13. 如有必要，修整凝胶加样孔、边缘及底部，并将其滑入印迹盘中。从凝胶下方用戴手套的手指去除气泡。

14. 将正确标记的膜置于相应的凝胶上面，凝胶左上角具有切口标记。去除气泡。如果凝胶上有一组样品，则在膜边缘进行标记，待转移后将其切除。

15. 在每块凝胶的膜上，都用另一块已用 10×SSC 润湿的 3MM Whatman 滤纸进行覆盖。去除气泡。然后将其暴露部分用条形石蜡膜或 Saran 带覆盖。注意避免覆盖含 DNA 的凝胶条带。将一叠干燥的支撑物（约 2cm 高）铺在 3MM Whatman 滤纸上，然后再在其上铺上纸巾（2.54～5.08cm 高）。

16. 放置一个凝胶固定器并施加约 500g 重量于纸巾顶部（如用一瓶 500ml 的水）。再把 10×SSC 倒入印迹盘中，距湿润的印迹层不到 1.27cm（0.5 英寸），使 DNA 持续转移 8～24 小时。纸巾湿润后需要更换。尽量防止整堆纸巾被缓冲液润湿。

17. 移去凝胶上方的纸巾和印迹纸张。将凝胶和附着其上的膜翻转过来平铺于干燥的印迹纸上，凝胶面向上。用非常软的铅笔或圆珠笔在膜上标记凝胶槽的位置。

18. 将凝胶从膜上剥离，弃去凝胶。

19. 室温下将膜浸在 6×SSC 中 5 分钟。

20. 对已转移到不带电荷的膜上的 DNA 进行固定。如果要用微波炉烘烤加以固定，首先将湿膜置于一张干燥的印迹纸上，然后用微波炉全火（750～900W）加热 2～3 分钟。

21. 将已固定的 DNA 直接与探针杂交。

（二）放射性标记的探针与固定于膜上的核酸进行 Southern 杂交

1. 将含目的 DNA 的膜漂浮在 6×SSC（或 6×SSPE）中，使膜从下向上完全润湿，并浸没 2 分钟。

2. 为了在热封袋中进行杂交，按照以下步骤对膜进行预杂交。①将湿润的膜滑入热封口袋中（如热封式塑料袋或类似物），按每平方厘米的膜加入 0.2ml 的比例加入预杂交溶液。尽量挤除袋中的空气。②用加热封口器封口并做第二次封口。轻轻挤压以检查袋子的强度及密封情况。将袋子浸泡于适当温度的水浴中，温育 1～2 小时（水溶液用 68℃；含 50% 甲酰胺的溶液用 42℃，含磷酸-SDS 的溶液用 65℃）。

3. 如果放射性标记的探针是双链 DNA，则将其加热到 100℃ 维持 5 分钟以使其变性。将探针置于冰水浴迅速降温。

4. 按以下方法进行探针与基因组 DNA 印迹杂交。①迅速将含有膜的袋子从水浴中取出。用剪刀从一角剪开袋子，倒入预杂交溶液。②将变性的探针加入适量的新鲜预杂交溶液中，再将该溶液倒入袋中。尽量除去袋中空气。③用热封口器再封口，保证尽量少的气泡留在袋中。为防止水浴的放射性污染，将二次封口的袋子封于未污染的袋中。在适当温度的水浴中温育袋子，保证所需要的杂交时间。

5. 杂交后洗膜。①戴手套，从水浴中取出袋子，去除外袋，迅速将内袋从一角切开。杂交溶液倒入用于抛弃放射性物质的容器中，然后沿袋子 3 个边切开袋子。②于室温除去膜，迅速将其浸泡于盛有数百毫升含 2×SSC 和 0.5% SDS 的盘中（约 1ml/cm^2 膜）。在缓慢振荡器上轻轻振摇盘子。

6. 5 分钟后，将第一次洗涤的溶液倒入放射性抛弃容器中，在盘中再加入数百毫升含 2×SSC 和 0.1% SDS 的溶液。室温温育并间歇轻微振荡 15 分钟。

7. 用数百毫升新鲜的含 0.1×SSC 和 0.1% SDS 的溶液替换洗涤溶液。膜于 65℃温育 30 分钟至 4 小时并轻微振荡。

8. 室温用 0.1×SSC 溶液对膜进行简单漂洗。

9. 将膜放置于纸巾上以除去膜上大部分液体。将湿膜置于一叠 Saran Wrap 纸上。在 Saran Wrap的对称位置用放射性墨水（或荧光点）做点标记。这些标记有助于放射自显影与膜进行比对。用 Scotch 带覆盖标记处，以防止胶片盒或增感屏被放射性墨水污染。

10. 用一张 Saran Wrap 纸盖膜，在 −70℃下将膜暴露于有增感屏的 X 线片上曝光 16 ~ 24 小时，以获得放射自显影图像。

Experiment 10　Southern blotting

Introduction

Southern blotting was the first analytical technique based on nucleic acid hybridization, which was invented by Edward Southern of the University of Oxford. This technique is mainly used to detect DNA molecules and to analyze the structure of genes.

Purpose

1. Detect the presence or absence of specific gene fragments in a complex mixture.
2. Determine the length of DNA restriction fragments harboring specific genes.

Principle

The first step in Southern blotting involves digestion of the DNA with a restriction enzyme. The resulting DNA fragments are then separated by an agarose gel electrophoresis. After electrophoresis, the gel is photographed to record the migration of DNA molecular weight markers, then soaked in an alkaline solution that denatures the fragments making them suitable for hybridization. The second step involves transferring the denatured fragments out of the gel and onto a nylon membrane (or sometimes nitrocellulose) where they become accessible for analysis using a probe. The gel is placed on a buffer-saturated spongey surface, the membrane positioned on top of the gel, and dry blotting papers are added on top of the membrane. The buffer moves by capillary action through the gel and into the papers above. The denatured DNA in the gel moves passively with the buffer and is stopped by the membrane. DNA is permanently bound to the membrane by baking or cross – linking with UV. By blotting, the DNA is bound to the membrane in the same position (relative to other fragments) as in the gel. The third step involves incubating the membrane with a labeled probe. This is known as hybridization and is carried out at a temperature and in a buffer that favor the formation of hybrids between the probe and fragments of DNA bound to the membrane. The membrane is then washed with a buffer to remove probe that is bound nonspecifically so that only labeled probe bound to target sequences remains. If the probe is labeled with radioactivity, the nylon membrane can be analyzed by autoradiography. Development of the X – ray film will reveal one or more dark bands which correspond to thc position of fragments bound by the probe. In this way the structure of individual genes can be characterized in terms of the number of hybridizing fragments and their sizes.

Reagents and apparatus

1. Nucleic acids and oligonucleotides.
2. Genomic DNA.
3. Salmon sperm DNA (approx. 10mg/ml).
4. Radiolabeled probe DNA or RNA (sp. act. 10^9cpm/μg).
5. Restriction enzymes.
6. 1 × TAE buffer.
7. 3M NaCl.
8. 10N NaOH.
9. 1M Tris – HCl, pH 8.0.
10. 0.7% 1 × TAE agarose.
11. 10% SDS.
12. 20 × SSC (20L): Dissolve 3504g NaCl and 1760g sodium citrate in 10L ddH_2O (Use a 20L carboy). Adjust the volume to 20L with ddH_2O, and the pH to 7.4 with several drops of concentrated HCl.

Procedures

1. Blotting

(1) Digest an appropriate amount of genomic DNA with one or more restriction enzymes.

(2) If necessary, concentrate the DNA fragments at the end of the digestion by ethanol precipitation. Dissolve the DNAs in approx. 25μl of TE (pH 8.0).

(3) Measure the concentrations of the digested DNAs by fluorometry or by the ethidium bromide or SYBR Gold spot test. Transfer the appropriate amount of each digest to a fresh microfuge tube. Add 0.15 volume of 6 × sucrose gel – loading buffer and separate the fragments of DNA by electrophoresis through an agarose gel (for most genomic DNAs, a 0.7% gel cast in 0.5 × TBE or 1 × TAE may be used). Maintain a low voltage through the gel (< 1V/cm) to allow the DNA to migrate slowly (If the digested DNAs have been stored at 4℃, they should be heated to 56℃ for 2 – 3minutes before they are applied to the gel. This heating disrupts any base pairing that may have occurred between protruding cohesive termini).

(4) After electrophoresis is complete, stain the gel with ethidium bromide or SYBR Gold and photograph the gel. Place a transparent ruler alongside the gel so that the distance that any band of DNA has migrated can be read directly from the photographic image.

(5) After fractionating the DNA by gel electrophoresis, transfer the gel to a glass baking dish. Use a razor blade to trim away unused areas of the gel, including the section of gelabove the wells. Be sure to leave enough of the wells attached to the gel so that the positions of the lanes can be marked on the membrane after transfer of DNA. Cut off a small triangular piece from the bottom left – hand corner of the gel to simplify orientation during the succeeding operations. Cut off the lanes containing the molecular – weight markers.

(6) To denature the DNA in the gels, by shaking gels slowly in a freshly prepared solution of 0.6M NaCl, 0.2N NaOH (200ml 3M NaCl + 20ml 10N NaOH brought to 1000ml with ddH_2O) for 30minutes. Check frequently to be certain the gels stay immersed in the denaturant. (A common problem is that if the gel surface remains dry, the DNA will not denature properly resulting in poor transfer. Often the gels can be kept submerged with a sunken clean glass test tube positioned on top of the gel. Rinse gels briefly in ddH_2O).

(7) To neutralize the gels, shake slowly in 1.5M NaCl, 0.5M Tris – HCl pH 8.0 (for 1 liter: 500ml of 3M NaCl + 500ml of 1M Tris – HCl). Check frequently to be certain gels stay immersed.

(8) If using prepared Southern set – ups (Fig. 10 – 1), remove the top blot blocks from each and replace them with ones freshly soaked in 10 × SSC. If a new transfer set – up is needed, wet blot blocks one by one by soaking them briefly in 10 × SSC and transferring them to a fresh blotting tray. Layer blot blocks to a thickness of about 1.27cm. When layering the blot blocks, gently roll a clean test tube across the top to remove air bubbles between blocks.

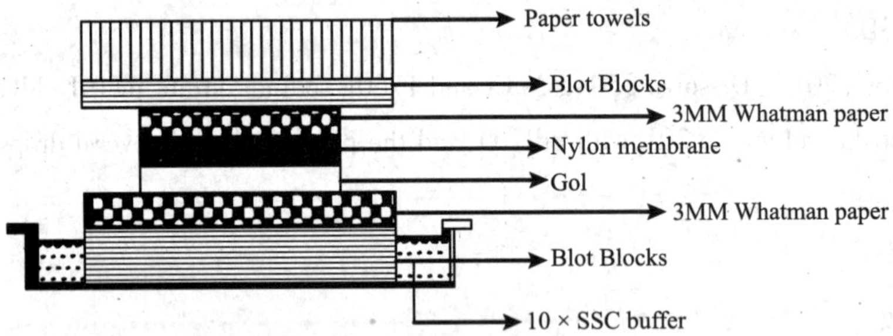

Fig. 10 – 1 set – up for Southern Blotting

(9) Fill the trays with 10 × SSC buffer up to the top of the stack (but do not immerse the top of the stack).

(10) Cut 3MM Whatman papers (or use pre – cut 3MM Whatman paper) to the cize of tho blot blocks, individually wet them in 10 × SSC, and lay one on each stack of blot blocks. Remove air bubbles by rolling the test tube as above.

(11) Cut Zetabind (or any other nylon membrane used as a solid support) slightly larger than the gels with a pair of clean scissors (wear gloves when handling membranes and protect membranes by working on clean 3MM paper). Label the membranes across the top with the respective I. D. #, the enzyme and the blot number using a dry erase marker and cut across the left hand top comer of

the membranes for orientation purposes.

(12) Wet the nylon membranes briefly in ddH_2O, transfer to a tray containing 10 × SSC and soak for 20minutes.

(13) Trim the gels at the wells, the bottom and sides (if necessary) and slide onto the blotting tray. Remove any air bubbles from under the gels with gloved fingers.

(14) Place the correctly labeled membrane on top of each corresponding gel with the notch oriented towards the top left hand corner. Remove air bubbles. If more than one family is on a gel, mark the membrane edges where cuts will be made after the transfer is complete.

(15) For each gel, wet another piece of 3MM Whatman paper (cut to the size of the gel) in 10 × SSC and place it on the membrane. Remove air bubbles. Cover the exposed portion of the wet blot blocks with strips of parafilm or Saran wrap (which act as a wicking barrier). Be certain not to cover any gel lanes containing DNA. Place a stack of dry blot blocks (about 2cm high) on top the 3MM Whatman paper and place a set of paper towels (2.54 – 5.08cm high) on top.

(16) Place a gel restrainer and a weight of about 500g on top of each stack of paper towels (a 500ml bottle of water is approximately 500g). Refill the blotting trays with 10 × SSC to within 1.27cm from the top of the wet blot blocks and allow the transfer of DNA to proceed for 8 – 24hours. Replace the paper towels as they become wet. Try to prevent the entire stack of towels from becoming wet with buffer.

(17) Remove the paper towels and the blotting papers above the gel. Turn the gel and the attached membrane over and lay them, gel side up, on a dry sheet of blotting paper. Mark the positions of the gel slots on the membrane with a very soft lead pencil or a ballpoint pen.

(18) Peel the gel from the membrane and discard the gel.

(19) Soak the membrane in 6 × SSC for 5minutes at room temperature.

(20) Immobilize the DNA that has been transferred to uncharged membranes. To fix by baking in a microwave oven. ①Place the damp membrane on a dry piece of blotting paper. ②Heat the membrane for 2 – 3minutes at full power in a microwave oven (750 – 900W).

(21) Proceed directly to hybridization of immobilized DNA to a probe.

2. Southern hybridization of radiolabeled probes to nucleic acids immobilized on membranes

(1) Float the membrane containing the target DNA on the surface of a tray of 6 × SSC (or 6 × SSPE) until the membrane becomes thoroughly wetted from beneath. Submerge the membrane for 2minutes.

(2) For hybridization in a heat – sealable bag. Prehybridize the membrane by the following methods. ①Slip the wet membrane into a heat – sealable bag (e.g., Sears Seal – A – Meal or equivalent), and add 0.2ml of prehybridization solution for each cm^2 of membrane. Squeeze as much air as possible from the bag. ②Seal the open end of the bag with a heat sealer and then make a second seal. Test the strength and integrity of the seal by gently squeezing the bag. Incubate the bag for 1 – 2hours submerged in a water bath set to the appropriate temperature (68℃ for aqueous solvents; 42℃ for solvents containing 50% formamide; 65℃ for phosphate – SDS solvents).

(3) If the radiolabeled probe is double – stranded DNA, denature it by heating for 5minutes at

100℃. Chill the probe rapidly in ice water.

(4) To hybridize the probe to a blot containing genomic DNA, carry out the following methods. ①Working quickly, remove the bag containing the membrane from the water bath. Open the bag by cutting off one corner with scissors and pour off the prehybridization solution. ②Add the denatured probe to an appropriate amount of fresh prehybridization solution and deliver the solution into the bag. Squeeze as much air as possible from the bag. ③Reseal the bag with the heat sealer; make sure that as few bubbles as possible are trapped in the bag. To avoid radioactive contamination of the water bath, seal the resealed bag inside a second, noncontaminated bag. Incubate the bag submerged in a water bath set at the appropriate temperature for the required period of hybridization.

(5) After hybridization, wash the membrane. ①Wearing gloves, remove the bag from the water bath, remove the outer bag, and immediately cut off one comer of the inner bag. Pour out the hybridization solution into a container suitable for disposal of radioactivity, and then cut the bag along the length of three sides. ②Remove the membrane and immediately submerge it in a tray containing several hundred ml of 2 × SSC and 0.5% SDS (i. e. approx. 1ml/cm^2 membrane) at room temperature. Agitate the tray gently on a slowly rotating platform.

(6) After 5minutes, pour off the first rinse solution into a radioactivity disposal container and add several hundred milliliters of 2 × SSC and 0.1% SDS to the tray. Incubate for 15minutes at room temperature with occasional gentle agitation.

(7) Replace the rinse solution with several hundred milliliters of fresh 0.1 × SSC with 0.1% SDS. Incubate the membrane for 30minutes to 4hours at 65℃ with gentle agitation.

(8) Briefly wash the membrane with 0.1 × SSC at room temperature.

(9) Remove most of the liquid from the membrane by placing it on a pad of paper towels. Place the damp membrane on a sheet of Saran Wrap. Apply adhesive dot labels marked with radioactive ink (or phosphorescent dots) to several asymmetric locations on the Saran Wrap. These markers serve to align the autoradiograph with the membrane. Cover the labels with Scotch Tape. This prevents contamination of the film holder or intensifying screen with the radioactive ink.

(10) Cover the membrane with a sheet of Saran Wrap, and expose the membrane to X－ray film for 16－24hours at －70℃ with an intensifying screen to obtain an autoradiographic image.

实验十一　DNA 序列测定

扫码“学一学”

【导言】

目前应用的序列测定技术基于两种实验方法：Sanger 双脱氧链末端终止法以及 Maxam 和 Gilbert 发展的化学降解方法。由于链终止法更有效且实施起来较为简单，现在已经替代了化学降解方法。

【目的】

了解 DNA 序列测定技术。

【原理】

双脱氧链末端终止方法利用 DNA 聚合酶延伸引物的能力，加入脱氧核苷三磷酸（dNTP）与被测序的模板杂交，直到链终止核苷酸掺入为止。DNA 序列测定，必须建立 4 组独立的酶促反应。4 组独立的反应包括：单链模板 DNA、DNA 聚合酶、引物、4 种脱氧核苷三磷酸中的一种（用于 DNA 合成）和少量双脱氧核苷三磷酸（ddNTPs）。在反应中，因为 ddNTPs 缺乏链延伸所必需的 3′-OH 基团，相应的双脱氧核苷酸使正在延伸的 DNA 链随机分别终止于模板链上的每一个 A，C，G，T 位置上。dNTP 和 ddNTP 随链的延伸和终止，随机且较均匀地产生数百至数千碱基对长度的片断。产生的片断都有相同的起点和不同的终点，借由高分辨率变性聚丙烯酰胺凝胶电泳分离。将放射性的磷或硫标记的 dNTP 加入测序反应，合成的 DNA 带有放射性标记，聚丙烯酰胺凝胶分离后，用 X 线照射显影，称为放射自显影技术。X 线胶片冲洗之后在 4 个泳道上会显示出阶梯状条带。模板 DNA 的序列能借由凝胶电泳每个泳道中最近的条带到较远的条带连续读出（图 11-1）。

一些不同的 DNA 聚合酶均可以用于测序反应，包括大肠埃希菌的 DNA 聚合酶的片断（称为 Klenow 片断），来自于噬菌体 T_7 的经过遗传修饰的 DNA 聚合酶（称为测序酶）和聚合酶链反应中所用的 *Taq* DNA 聚合酶。

在以单链 DNA 为模板的合成中，所有的 DNA 聚合酶都要求短的双链 DNA 区域作为起始。这可以通过在测序反应中加入短的单链 DNA（化学合成的寡核苷酸引物）实现。选择的引物序列与模板 DNA 互补，形成一个短的双链区域，以决定测序反应的起始。

双脱氧链末端终止法测序的一个要求是 DNA 必须以单链形式存在，以便 DNA 聚合酶结合并复制。最初是将 DNA 克隆到噬菌体载体 M13 中实现，当重组噬菌体转染大肠埃希菌时，产生测序模板的单链拷贝，可用作测序模板。近来用于测序的单链 DNA 模板可以用更简单的方法获得，即采用热或碱变性的方法而无须进行克隆。

A 部分　双脱氧链终止法测序用变性模板的制备

在该实验过程中，双链质粒 DNA 碱变性，与适当的寡核苷酸引物退火结合。

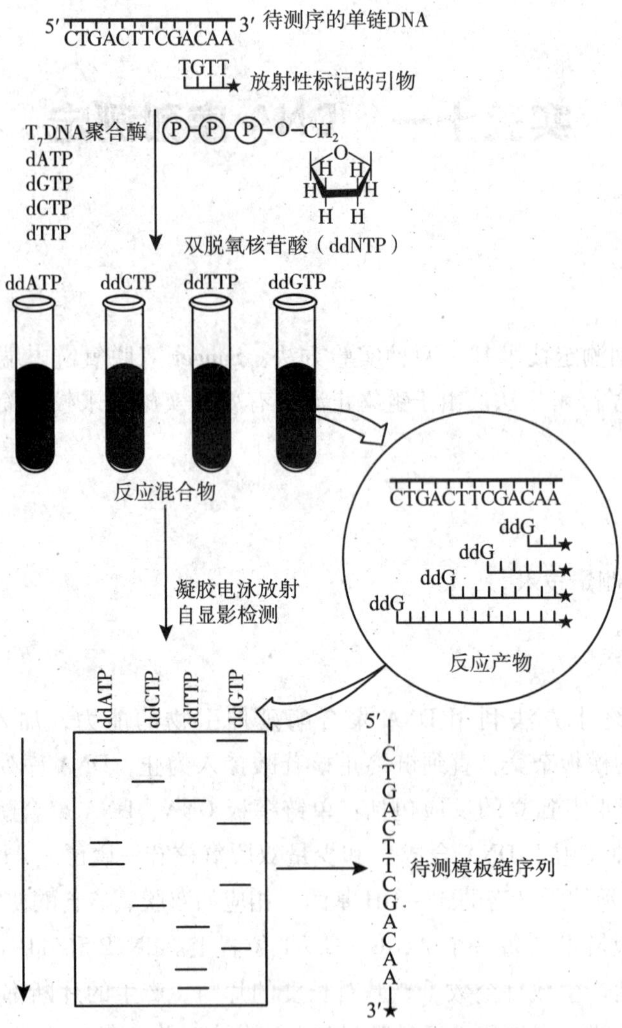

图 11-1　双脱氧链末端终止法 DNA 测序原理

【试剂与器材】

1. 醋酸铵（5M，未加缓冲液，pH 约 7.4）。
2. 氯仿：异戊醇（24：1，*V/V*）。
3. DMSO。
4. NaOH（2N）/EDTA（2mmol）。
5. 酚，被 H_2O 或 TE 平衡（pH 7.6）。
6. 测序酶反应缓冲液。
7. 核酸和寡核苷酸：①寡核苷酸引物（20～30 个核苷酸）。②闭合环状双链 DNA。

【实验步骤】

1. 将约 5μg 纯化质粒 DNA 转移至 1.5ml 的微离心管中。补加水至体积为 50μl。加入 10μl 新配的 2N NaOH/2mM EDTA 溶液。室温孵育 5 分钟。

2. 加入 5M 醋酸铵 30μl，漩涡混匀管中的溶液。

3. 加入平衡酚 45μl，漩涡混匀管中的溶液。

4. 加入氯仿：异戊醇 135μl，漩涡混匀管中的溶液，然后在室温下以最大的速度离心

15 分钟使之分层。

5. 小心吸取刚离心完的上层水相至一个新的微离心管中，加入冰预冷的乙醇 330μl 沉淀 DNA。混匀管中的溶液，放置于干冰 - 乙醇浴中 15 分钟。

6. 在 4℃以最大的速度离心 15 分钟以回收变性的 DNA。

7. 非常小心地轻轻倒出上清液，不要搅动 DNA 沉淀（该沉淀浮在管边上可能看见也可能看不见）。

8. 将管内溶液再次离心 2 秒，尖嘴 Tip 头尽量除去痕量乙醇，避免接触 DNA 沉淀。

9. 将沉淀在室温下干燥，再次悬浮于水中使其浓度为 1μg/μl。

10. 将引物 DNA 与变性的模板进行退火，碱变性的质粒 DNA 5μg（5μl）平均分配到一组 4 个反应管中（1.25μg 变性 DNA 加入 ddA 反应管中，1.25μg 变性 DNA 加入 ddG 反应管中，以此类推）。按以下配方建立退火反应：

碱变性 DNA	5.0μg（于 5μl 水中）
DMSO（任选）	2.0μl
测序酶反应缓冲液	2.0μl
测序引物	1.0μl（10ng）
水	至 11.0μl

65℃加热混合物 2 分钟，然后慢慢冷却至室温。将退火后的样品保存于冰上，直至进行链延伸终止反应。

B 部分　使用噬菌体 T_7DNA 聚合酶（测序酶）进行双脱氧链终止反应

该方法描述了以单链 DNA 为模板用测序酶进行催化的测序反应（2.0 版，USB/Amersham Life Science）。

【试剂与器材】

1. 缓冲液和溶液

（1）去离子蒸馏水（冰预冷）。

（2）含 4 种 ddNTP 中一种的溶液（0.5mM）。

（3）含 4 种 dNTP 中一种的溶液（0.5mM）。

（4）二硫苏糖醇（DTT）（100mM）。

（5）5 × 标记混合物和 ddNTP 链延伸终止混合物（ddCTP，ddTTP，ddATP 和 ddGTP）。这些反应混合物按表 11 - 1 中所显示的溶液体积进行配比制备。

表 11 - 1　测序反应中 5 × 标记混合液和 ddNTP 延伸/终止混合液

反应混合物	dNTP 和 ddNTP 储存液/微升								其他试剂	
	dCTP 0.5mmol/L	dTTP 0.5mmol/L	dATP 0.5mmol/L	dGTP 0.5mmol/L	ddCTP 0.5mmol/L	ddTTP 0.5mmol/L	ddATP 0.5mmol/L	ddGTP 0.5mmol/L	NaCl 0.5mmol/L	H_2O
标记	15	15	—	15	—	—	—	—	—	955
ddCTP	160	160	160	160	16	—	—	—	10	334
ddTTP	160	160	160	160	—	16	—	—	10	334
ddATP	160	160	160	160	—	—	16	—	10	334
ddGTP	160	160	160	160	—	—	—	16	10	334

（6）甲酰胺上样缓冲液。

（7）测序酶稀释缓冲液。

（8）5×含 $MgCl_2$的测序酶反应缓冲液。

（9）5×含 $MnCl_2$的测序酶反应缓冲液。

（10）TE（pH 7.6）。

2. 酶

（1）测序酶（版本 2.0）。由制造商提供，约 13 单位/微升（约 1mg/ml）。-20℃储存（无需使用无霜冰箱）。

（2）酵母无机焦磷酸酶。

3. 核酸和寡核苷酸

（1）寡核苷酸引物（用水或 TE 溶解，约 3.3ng/μl，pH 7.6）。

（2）模板 DNA（1μg，溶于 TE 中）（pH 7.6）。

4. 放射性混合物　［α-^{32}P］dATP（1000Ci/mmol，10mCi/ml）。

【实验步骤】

1. 当引物和模板冷却时，融化 5×标记、ddNTP 链延伸终止混合物和放射性标记的 dATP。将融化的溶液储存于冰上。

2. 分别取 2.5μl 各种 ddNTP 链延伸终止混合物加至不同的 0.5ml 微量离心管中或加入微量滴定板的微孔中，分别标上不同的颜色或 C，T，A 和 G。

3. 用冰预冷的水将 5×标记混合物稀释 5 倍。每个测序反应需要 2μl 稀释过的标记混合物。

4. 用冰预冷的测序酶稀释缓冲液稀释测序酶，加或不加酵母无机焦磷酸酶。每个测序反应管中的模板需要体积为 2.0μl、约 3.0 单位的测序酶。稀释的酶需储存于冰上。

5. 进行标记反应时，需要将以下组分加入 10μl Part A 的退火反应体系中：

稀释后的标记混合物（第 3 步骤）	2μl
0.1M 二硫苏糖醇	1μl
［α-^{32}P］ATP	0.5μl
稀释后的测序酶（约 1.6 单位/微升）	2.0μl

轻轻敲打管壁或微量滴定板以混合反应组分，然后于 20℃孵育 2～5 分钟。稀释后的测序酶储存于冰上，不可放置于室温。

6. 在标记反应即将结束时，于 37℃预热微量离心管或微量滴定板建立终止反应。这一步很重要！然后转移 3.5μl 标记混合物至预热并已做标记的含有适量双脱氧终止混合物的微量离心管管壁上或预热的含有适量双脱氧终止混合物的微量滴定板的微孔边上（上述第 2 步）。

7. 室温下将微量离心管置于微型离心机中（使用 0.5ml 离心管的转子或衔接头，或放入已剪去头部的 1.5ml 管中），或室温下将微量滴定板置于有合适衔接头的离心机中。将 C，T，A 和 G 各管或板于 2000 转/分离心数秒从而混合其中的反应组分。然后立刻转移至 37℃加热器上或水浴 3～5 分钟。

8. 加入甲酰胺上样缓冲液 4μl 终止反应。

9. 反应物在 -20℃可储存 5 天或直接用变性凝胶电泳进行分析。热变性（100℃，2 分钟）之后迅速在冰上进行冷却，分别取3μl C，T，A 和 G 反应物加到测序凝胶的加样孔中。

C 部分　变性聚丙烯酰胺凝胶的制备

1978 年，Fred Sanger 和 Alan Coulson 发明了一种铺制薄型聚丙烯酰胺凝胶并进行电泳的方法，现已作为检测 DNA 测序反应结果的常规方法。

【试剂】

1. 丙烯酰胺溶液（45%，*W/V*）。
2. 过硫酸铵（1.6%，*W/V*）。
3. 去离子水。
4. 去污剂（家用清洁剂）。
5. 乙醇。
6. KOH/甲醇溶液。
7. 硅烷化溶液：Gel Slick（FMC Bioproducts），RainX（Unelko，Scottsdale Arizon）和 Acrylease（Stratagene）。
8. 10×TBE 电泳缓冲液。
9. TBE。
10. *N*，*N*，*N′*，*N′*-四甲基乙二胺（TEMED）。
11. 尿素，固体。

【实验步骤】

1. 如有必要，用刷子蘸 KOH/甲醇溶液拭去板上旧的硅烷化试剂。为避免皮肤油脂污染玻板表面，在处理玻板时需戴无滑石粉的手套，且只能接触玻板边缘。
2. 用温热的稀释过的清洁液清洗玻板和间隔片，然后用自来水彻底冲洗，再用去离子水冲洗。最后用无水乙醇冲洗玻板除去水分，挥发干燥。
3. 用硅烷化溶液处理较小的或带槽口的玻板的内表面。在化学通风橱中，将玻板内表面朝上放置于一块纸巾上，在玻板的表面倒上或喷上少量硅烷化试剂。用 Kimwipes 纸擦掉整个表面上的液体，在空气中干燥（1~2 分钟）。先用去离子水冲洗玻板，然后用乙醇冲洗，挥发干燥。
4. 将比较大的（或无槽口）玻板放在空的试管架上（清洁的一面向上），将两块间隔片沿玻板边缘放置妥当。
5. 将较小的（刻槽口）玻板的已硅烷化的一面向下，放置于大的（无槽口）玻板的上面，同时保证间隔片位于两块玻板的边缘位置。
6. 用2~3 个大弹簧夹（5cm 长）将玻板的一边夹起，并将玻板的另一边以及底部用密封带封好。
7. 除去弹簧夹并将这一边用密封带封好。
8. 将鲨鱼齿梳的平齐端放入胶模敞开口的一侧，检查是否合适。然后取下梳子，并将空的凝胶模放置在试管架上。
9. 在实验台上的工作区铺上有塑料支撑的保护纸。

10. 在一个 250ml 的锥形瓶中，按表 11 - 2 配制所需浓度的含丙烯酰胺的测序凝胶溶液。表 11 - 2 给出的凝胶体积足以铺制一块 40cm × 40cm 的测序凝胶，且可以根据所需凝胶的大小按比例调整。凝胶的配制必须一次完成，中间不能间断。

表 11 - 2　变性胶溶液配方

	4%胶	6%胶	8%胶	10%胶
丙烯酰胺溶液	8.9ml	13.3ml	17.8ml	22.2ml
10 × TBE 缓冲液	10.0ml	10.0ml	10.0ml	10.0ml
水	45.8ml	41.4ml	36.9ml	32.5ml
尿素	42.0g	42.0g	42.0g	42.0g

11. 混合所有试剂，于 55℃水浴加热 3 分钟以促进尿素溶解。

12. 将溶液从水浴取出，于室温冷却 15 分钟。不时旋转混合物。

13. 将锥形瓶中的溶液用真空泵抽真空。

14. 将溶液移至 250ml 的玻璃烧杯中，加入 3.3ml 新配制的 1.6% 过硫酸铵，缓慢旋转混匀。

15. 加入 50μl TEMED 溶液，缓慢旋转混匀。直接将凝胶溶液从烧杯倒入凝胶模内。也可以用 60ml 皮下注射器吸取约 40ml 的胶溶液。不要将空气吸入注射器内！因为凝胶溶液很快聚合，因此操作时应尽量加快速度。将凝胶溶液置于冰上可适当减慢聚合速度，如操作者经验不足可采用该方法。

16. 用手托起胶模使之与水平线约呈 45°，将烧杯或注射器中的凝胶溶液从胶模的顶部角落位置缓慢注入，同时保持胶模与水平线之间呈 45°。

17. 将胶模放置于试管架上。

18. 立即将鲨鱼齿梳平整的一侧插入凝胶溶液内约 0.5cm。当将凝胶竖起时，梳子的两端插入液面的深度应相等。

19. 用大弹簧夹夹紧梳子插入的位置，使皮下注射器或吸量管中剩余的丙烯酰胺/尿素溶液沿凝胶的顶端形成一串丙烯酰胺液滴。凝胶在室温至少聚合 15 分钟。

20. 洗净 60ml 的注射器，避免注射器因为其中的丙烯酰胺聚合而堵塞。

21. 聚合 15 分钟后对凝胶进行检查，若聚合完全则在梳子平整面的下面出现 Schlieren 线。聚合完全后（在倒入凝胶后约 1 小时），取下大弹簧夹。

22. 凝胶可立刻使用，也可于室温保存 24 小时或 4℃保存 48 小时再用。为防止保存期间发生脱水现象，可将梳子留于凝胶内，并在凝胶顶端周围放一些 1 × TBE 的润湿纸巾，可用 Saran 保鲜膜覆盖。在此阶段不要拔掉梳子。

D 部分　DNA 测序凝胶的加样和电泳

确认所使用的电泳槽绝缘良好，盛放缓冲液的槽已加盖，且放置于干燥稳定的实验台上。加样或拆胶之前必须关掉电源。

【试剂】

1. 5 × TBE。

2. 1×TBE。

3. DNA 测序反应物。

【实验步骤】

1. 用湿润的纸巾或海绵拭去玻板外侧的聚丙烯酰胺/尿素，用吸量管吸数毫升 1×TBE 缓冲液加到比较小的或带槽口的玻璃的顶部，缓慢将梳子从凝胶中取出。用刀片把绝缘胶片从底部切开并从胶模底部撕去。不可将胶带从凝胶边上撕开！

2. 按厂家说明用弹簧夹、塑料夹或电泳自带的螺纹夹将胶模固定在电泳装置上。比较小的或带槽口的玻板应直接与电泳装置接触。比较大的玻板则应朝向实验人员。

3. 将电泳装置的上下槽装满合适的电泳缓冲液。

（1）将上槽和下槽装满 1×TBE。确保下槽中的缓冲液液面恰好淹没板的底部，而上槽中缓冲液的液面应该恰好在比较小的或带槽口的板的上沿，且直接与凝胶接触。

（2）用 10ml 注射器装满 1×TBE 冲洗凝胶的顶端，将剩余的聚丙烯酰胺及尿素从凝胶中去除，如果需要可用注射器针头刮掉黏附在玻板上的聚丙烯酰胺。用 Pasteur 吸量管去除下槽玻璃板底部的气泡。

（3）将电极与电泳装置以及电源装置相连。负极（黑色）应接到上槽，正极（红色）则接到下槽。接上固有的热传感器（如有）或温度 - 监视条。恒定功率（50～70W）条件下，电泳约 45 分钟，或直至凝胶温度达到 45～50℃ 为止，关掉电源，拆开电极。

4. 将含有测序反应物的微离心管在 100℃ 的加热装置中处理 2 分钟。如果反应在微孔板中进行，则只需除去盖子于 85℃ 水浴中漂浮 5 分钟即可。

5. 当加热处理试管或微孔板正在孵育时，可用有 22 号皮下注射针头的 10ml 注射器吸取适量的 0.5×TBE 或 1×TBE 溶液。用力推出其中的溶液冲洗液面下的用于加样的凝胶面以除去残留的尿素和聚丙烯酰胺凝胶碎片，直至在加样区域看不到残留的尿素。

6. 将鲨鱼齿梳轻轻插入（齿向下）加样槽内。向下推动梳子直到梳齿穿透凝胶表面。

7. 将微量离心管或微量滴定板从水浴或加热器中转移到冰上。加样前一直将样品保持在 0℃，骤冷是为了减少模板和放射性标记链之间的复性。

8. 取每种测序反应物 1～5μl 加入凝胶中相邻的上样槽中。

9. 当所有的样品都加样完毕时，将电极接到电源和电泳装置上，负极（黑色）接上槽，正极（红色）接下槽。电泳时维持一定的功率以确保凝胶温度为 45～50℃。凝胶大小与功率、电压的关系见表 11－3。

表 11－3　凝胶大小与功率、电压的关系

凝胶大小（cm）	功率（W）	电压（V）
20×40	35～40	1700
40×40	55～75	2000～3000
40×40（电解质梯度凝胶）	75	2100

依据甲酰胺上样缓冲液中的指示染料的迁移情况监控整个电泳过程（表 11－4）。一般二甲苯蓝染料迁移到距离底部 5～10cm，溴酚蓝已到底部时说明电泳已完毕。表 11－4 中数字（核苷酸）只表示 DNA 的粗略大小，在标准的测序凝胶中它们与指示染料迁移率相当。

表 11-4 变性的聚丙烯酰胺凝胶中标准参照染料的迁移率

聚丙烯酰胺凝胶（%）	溴酚蓝	二甲苯蓝 FF
5	35	130
6	29	106
8	26	76
10	12	55
20	8	28

10. 依据目的序列与寡核苷酸引物之间的距离，第一组样品中的溴酚蓝迁移至凝胶底部（需 1.5～2.0 小时）约 15 分钟后，可对标准或含甲酰胺的变性聚丙烯酰胺凝胶进行第二次加样。第一次加样所得到的序列离引物较远，而第二次加样所得到的序列离引物则较近。第一次上样后 2 小时再进行第二次上样，可使可读序列的长度延长约 35%。

（1）关掉电源并断开与序列测定装置之间的连接。

（2）更换上槽和下槽中的缓冲液。

（3）依照上述第 4 步的方法加热变性样品。

（4）加样。

（5）重新连接测序装置和电源。

（6）跑电泳，与第一次相同，维持一定的功率以确保凝胶温度为 45～50℃。

11. 电泳结束时，依照 Part E 所描述的方法取出凝胶，进行放射自显影。

E 部分 放射自显影以及测序胶的判读

【试剂】

1. 测序胶固定液。

2. 放射性墨水，可反复使用的化学发光标记物可以替代放射性墨水，可从 Stratagene（Glogos）购买。这种标记物可反复使用，只需在重新使用前于荧光灯下照射一下即可。

【实验步骤】

1. 电泳结束后，关掉电源并切断序列测定装置与电源之间的连接。倒弃电泳缓冲液，然后从电泳装置中取出胶模。

2. 将胶模平放在有塑料支撑的保护性实验台上，较小的（带槽口）板位于上方。等凝胶冷却到 37℃以下再进行后续操作。

3. 除去所有的凝胶密封带。用金属刮刀的末端缓慢地将玻板撬开。凝胶应仍粘连在比较长的（未硅烷化的）玻板上。

4. 玻板被分开后，在先加样的凝胶一侧的底部或顶端切下一个角，以便在后续操作中能够辨别凝胶的方位。

5. 将测序凝胶放在固定液（甲醇/醋酸）中进行固定。

6. 将凝胶连同支撑它的玻板一起转移至装有甲醇：醋酸固定液的浅盘中。室温固定 30 分钟。固定时不要搅动液体。

7. 30 分钟后，从固定液中非常小心地取出玻板和凝胶。尽量使玻板维持水平位置，直至大部分固定液沥干。将玻板（凝胶面向上）放置在一叠纸巾上。用 Kimwipes 纸吸去玻板上的剩余固定液，但勿使纸巾与凝胶表面接触。用戴手套的手指小心抹平凝胶皱折或破损处。

8. 准备一张 Whatman 3MM CHR 滤纸（或同等物），其长和宽均比凝胶略大（2 ~ 3cm）。先将滤纸稍卷一下，让其中心与凝胶中心相对，然后再逐渐将滤纸贴在凝胶的表面，轻压使凝胶与滤纸的粗糙面贴紧。

9. 一手将滤纸扶住，另一手将玻板提起，迅速将玻板翻转，放在一张干的保护纸上。小心移开上面的玻板，使得玻板与 3MM CHR 滤纸分离。当拿开玻璃板后，凝胶应与 3MM CHR 滤纸相贴。

10. 将 3MM CHR 滤纸（凝胶面向上）放置在相同大小的两张 Whatman 3MM 滤纸上。剪一张略长于且略宽于凝胶的 Saran 包装膜覆盖在凝胶上。注意勿其皱折及产生气泡。有他人帮忙，上述步骤较容易完成。将 Saran 包装膜的角向外拉以使其绷紧展平，然后将展平的 Saran 包装膜放到凝胶表面上。一旦包装膜接触凝胶，就不要再移动，否则容易撕坏凝胶。可用琼脂糖梳子、Kimwipes 纸或塑料卡的平端将膜与凝胶之间的气泡去除。

11. 用切纸刀或剪刀将 3 张 Whatman 纸和 Saran 包装膜修平，使其大小大致与凝胶相等。

12. 将滤纸、凝胶和 Saran 包装膜一起置于凝胶干燥机上，有 Saran 包装膜的一面向上。在干燥的时候，用一张 Mylar 纸保证凝胶平整。

13. 按照干燥机说明，在 80℃下将凝胶真空干燥 30 ~ 60 分钟。

14. 从干燥机中取出凝胶并剥去 Saran 包装膜，干燥的凝胶应触感平滑而不黏滞。如果发黏，一个快速的补救方法就是将乳胶手套翻转过来，用带有滑石粉的手套去拂凝胶。为了便于确定凝胶的方向，应在切去凝胶底角（步骤 4）后 3MM 滤纸上形成的空缺处贴上用放射性墨水或化学荧光墨水标记的黏性小标签纸。

15. 在暗房中，将干燥的凝胶（凝胶面向上）置于弹簧金属盒中。用一张未曝光的 X 线片覆盖凝胶，关上盒子。在室温或 -80℃曝光 16 ~ 24 小时，进行放射自显影。

16. 按照厂家的说明冲洗放射自显影片，读出 DNA 序列。

Experiment 11 DNA sequencing

Introduction

Two methods have been developed to determine the nucleotide sequence of DNA molecules: the dideoxy chain termination method of Sanger and the chemical degradation method developed by Maxam and Gilbert. The chain termination method has now superseded the chemical method because it is more efficient and is simpler to perform.

Purpose

To become familiar with DNA sequencing technology.

Principle

The dideoxy chain termination method is based on the ability of a DNA polymerase to extend a primer, hybridized to the template that is to be sequenced, until a chain - terminating nucleotide is incorporated. To sequence a DNA molecule four separate enzyme reactions are prepared. Each contains: template DNA in single - stranded form, DNA polymerase, primer, each of the four deoxynucleotide triphosphates (dNTPs) which are the building blocks for DNA synthesis and a limiting amount of a different modified nucleotide called a dideoxynucleotide triphosphate (ddNTP). Because ddNTPs lack the 3′ - OH group necessary for chain elongation, the growing oligonucleotide is terminated selectively at G, A, T or C, depending on the respective dideoxy analog in the reaction. The relative concentrations of each of the dNTPs and ddNTPs can be adjusted to give a nested set of terminated chains over several hundred to a few thousand bases in length. The resulting fragments, each with a common origin but ending in a different nucleotide, are separated according to size by high resolution denaturing polyacrylamide gel electrophoresis. By adding dNTP which contains radioactive phosphorus or sulfur to the sequencing reactions, the synthesized DNA becomes radioactive allowing it to be detected by placing the polyacrylamide gel against a sheet of X - ray film. This is called autoradiography. When the X - ray film is developed a series of bands is visible in the four lanes which form a ladder. The sequence of the template DNA can be determined by identifying the smallest band followed by successively larger bands and assigning the terminal base in each from its lane on the gel (Fig. 11 - 1). Several different DNA polymerases have been used for the sequencing reaction. These include part of the *E. coli* DNA polymerase I known as the Klenow fragment, a genetically modified DNA polymerase from the phage T_7, called Sequenase and *Taq* DNA polymerase which is also used in the polymerase chain reaction.

All DNA polymerases require short regions of double - stranded DNA to initiate DNA synthesis on a single - stranded template. This is provided in sequencing reactions by the addition of short, single - stranded DNA molecules called oligonucleotide primers which are produced by chemical synthesis. The sequence of the primer is chosen to be complementary to the template DNA such that it binds to it forming a short double - stranded region, thus determining the point at which the sequencing reaction is initiated.

A requirement for sequencing by the dideoxy chain termination method is that the DNA must be present in single - stranded form capable of being copied by a DNA polymerase. Previously this was achieved by cloning the DNA into the phage vector, M13. When recombinant phage were used to infect *E. coli*, single - stranded copies of the cloned sequence were produced which could be used as templates for sequencing. More recently, single stranded DNA for sequencing has been produced in a much simpler way without cloning by denaturation of the DNA template using heat or alkali.

Part A Preparing denatured templates for sequencing by dideoxy - mediated chain termination

In this protocol, double - stranded plasmid DNA is denatured by alkali and annealed to an appropriate oligonucleotide primer.

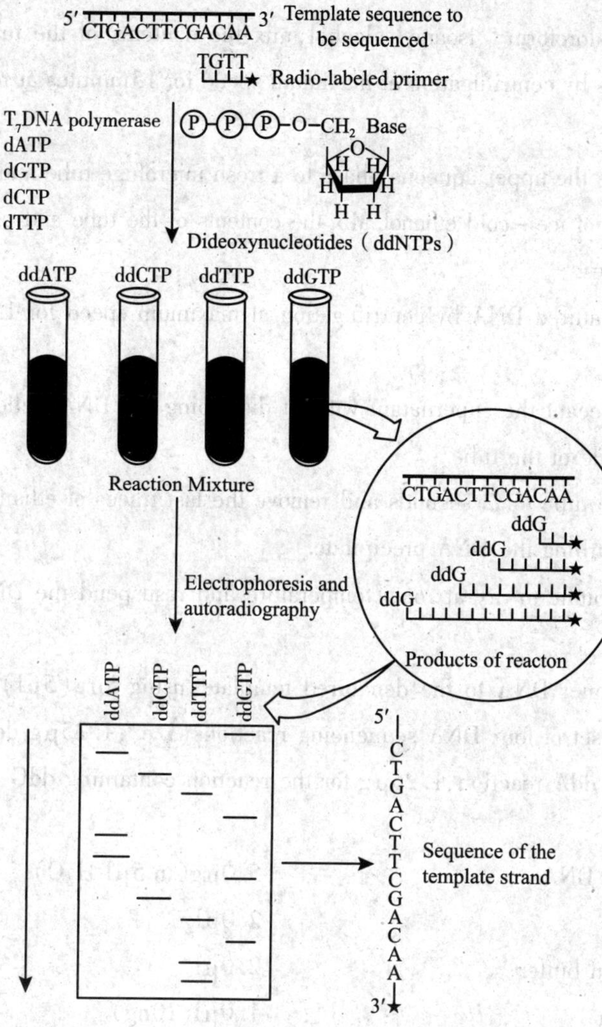

Fig. 11 - 1 Principle of DNA sequencing by Dideoxy - mediated Chain Termination

Reagents and Apparatus

1. Ammonium acetate(5M, unbuffered, pH approx. 7.4).
2. Chloroform : isoamyl alcohol(24 : 1, *V/V*).
3. DMSO.
4. NaOH(2N)/EDTA(2mM).
5. Phenol, equilibrated with H_2O or TE(pH 7.6).
6. Sequenase reaction buffer.
7. Nucleic acids and oligonucleotides: ①Oligonucleotide primers(20 - 30 nucleotides). ②Plasmid DNA, closed circular, double - stranded.

Procedures

1. Transfer approx. 5μg of purified plasmid DNA to a 1.5ml microfuge tube. Adjust the volume to 50μl with H_2O. Add 10μl of freshly prepared 2N NaOH/2mM EDTA solution. Incubate the mixture for 5minutes at room temperature.

2. Add 30μl of 5M ammonium acetate and mix the contents of the tube by vortexing.

3. Add 45μl of equilibrated phenol and mix the contents of the tube by vortexing.

4. Add 135μl of chloroform : isoamyl alcohol, mix the contents of the tube by vortexing, and then separate the phases by centrifugation at maximum speed for 15minutes at room temperature in a microfuge.

5. Carefully remove the upper aqueous phase to a fresh microfuge tube and precipitate the DNA by the addition of 330μl of ice - cold ethanol. Mix the contents of the tube and place it in a dry ice - ethanol bath for 15minutes.

6. Recover the denatured DNA by centrifugation at maximum speed for 15minutes at 4℃ in a microfuge.

7. Very carefully decant the supernatant without disturbing the DNA pellet, which may or may not be visible on the side of the tube.

8. Recentrifuge the tube for 2 seconds and remove the last traces of ethanol with a drawn - out pipette tip without disturbing the DNA precipitate.

9. Allow the precipitate to dry at room temperature and resuspend the DNA in H_2O at a concentration of 1μg/μl.

10. Anneal the primer DNA to the denatured template, using 5μg(5μl) of alkali - denatured plasmid DNA for each set of four DNA sequencing reactions(i. e. ,1. 25μg denatured DNA for the reaction containing the ddA reaction, 1. 25μg for the reaction containing ddG, etc.). Set up annealing reactions as follows:

alkali - denatured DNA	5. 0μg(in 5μl H_2O)
DMSO(optional	2. 0μl
Sequenase reaction buffer	2. 0μl
sequencing primer	1. 0μl(10ng)
H_2O	to 11. 0μl

Heat the mixture to 65℃ for 2minutes and then allow it to cool slowly to room temperature. Store the annealed samples on ice until the elongation/chain termination reactions are set up.

Part B Dideoxy - mediated Sequencing Reactions using bacteriophage T_7 DNA polymerase(sequenase)

This protocol describes DNA - sequencing of single - stranded DNA templates in reactions catalyzed by Sequenase(version 2. 0, USB/Amersham Life Science).

Reagents and apparatus

1. Buffers and solutions

(1) Deionized distilled H_2O(ice cold).

(2) ddNTP solution(0. 5mM) of each of the four ddNTPs.

(3) dNTP solution(0. 5mM) of each of the four dNTPs.

(4) Dithiothreitol(DTT) (100mM).

(5) 5 × Labeling mixture and ddNTP extension/termination mixtures(ddCTP, ddTTP, ddATP,

and ddGTP).

These reaction mixtures can be assembled by mixing the volumes of the solutions shown in table 11 - 1.

Table 11 - 1 5 × Labeling Mixture and ddNTP Extension/Termination Mixtures for Sequenase Reactions

Reaction Mixture	Stock Suolutions of dNTPs and ddNTPs(all volumes/μl)								Other Reagents	
	dCTP 0.5mmol/L	dTTP 0.5mmol/L	dATP 0.5mmol/L	dGTP 0.5mmol/L	ddCTP 0.5mmol/L	ddTTP 0.5mmol/L	ddATP 0.5mmol/L	ddGTP 0.5mmol/L	NaCl 0.5mmol/L	H_2O
Labeling	15	15	—	15	—	—	—	—	—	955
ddCTP	160	160	160	160	16	—			10	334
ddTTP	160	160	160	160	—	16	—	—	10	334
ddATP	160	160	160	160	—	—	16	—	10	334
ddGTP	160	160	160	160	—	—	—	16	10	334

(6) Formamide loading buffer.

(7) Sequenase dilution buffer.

(8) 5 × Sequenase reaction buffer with $MgCl_2$.

(9) 5 × Sequenase reaction buffer with $MnCl_2$.

(10) TE(pH 7.6).

2. Enzymes

(1) Sequenase(version 2.0).

Sequenase is supplied by the manufacturer at a concentration of approx. 13units/μl(approx. 1mg/ml). Store at -20℃(not in a frost - free freezer).

(2) Yeast inorganic pyrophosphatase.

3. Nucleic acids and oligonucleotides

(1) Oligonucleotide primer(approx. 3.3ng/μl in H_2O or TE, pH 7.6).

(2) Template DNAs(1μg) in TE(pH 7.6).

4. Radioactive Compounds

[α - ^{32}P]dATP(1000Ci/mmol, 10mCi/ml).

Procedures

1. While the primer and template are cooling, thaw the 5 × labeling and ddNTP extension/termination mixtures and radiolabeled dATP. Store the thawed solutions on ice.

2. Transfer 2.5μl of each ddNTP extension/termination mixture into separate 0.5ml microfuge tubes or into individual wells of a microtiter plate color - coded or labeled C, T, A, and G.

3. Make a fivefold dilution of the 5 × labeling mixture in ice - cold H_2O. A volume of 2μl of the diluted labeling mixture is required for each template sequenced.

4. Dilute Sequenase in ice - cold Sequenase dilution buffer, with or without addition of yeast pyrophosphatase. A volume of 2.0 containing approx. 3.0units of Sequenase enzyme, is required for each template sequenced. Store the diluted enzyme on ice at all times.

5. To carry out the labeling reaction, add the following to the 10μl annealing reaction of Part A:

diluted labeling mixture(from Step 3)	2μl
0.1M dithiothreitol	1μl
[α - ^{32}P]ATP	0.5μl
diluted Sequenase(approx. 1.6units/μl)	2.0μl

Mix the components of the reaction by gently tapping the sides of the tube or microtiter plate and then incubate the reaction for 2 - 5minutes at 20℃. Store the diluted Sequenase in ice; do not allow it to warm to ambient room temperature.

6. Toward the end of the labeling reaction, prepare to set up the termination reactions by prewarming the labeled microfuge tubes or microtiter plates to 37℃. This step is important! Then transfer the 3.5μl of the labeling reaction to the walls of each of the prewarmed labeled microfuge tubes or to the sides of prewarmed microtiter wells containing the appropriate dideoxy terminator mixtures (Step 2 above).

7. Place the microfuge tubes in a microfuge at room temperature(use appropriate rotor or adaptors for 0.5ml tubes or place inside decapitated 1.5ml tubes), or place the microtiter plates in a centrifuge at room temperature equipped with an appropriate adaptor. Centrifuge the C, T, A, and G tubes or plates for a few seconds at 2000rpm to mix the components of the reactions. Immediately transfer the reactions to a heating block or a water bath for 3 - 5minutes at 37℃.

8. Stop the reactions by adding 4μl of formamide loading buffer.

9. The reactions may be stored for up to 5days at - 20℃ or analyzed directly by denaturing gel electrophoresis. After heat denaturation(2minutes at 100℃) and quick cooling on ice, load 3μl of each of the C, T, A, and G reactions into individual wells of a sequencing gel.

Part C Preparation of denaturing polyacrylamide gels

In 1978, Fred Sanger and Alan Coulson devised a method to pour and run thin polyacrylamide gels, which are now used ubiquitously to resolve the products of DNA sequencing reactions.

Reagents and apparatus

1. Acrylamide solution(45%, *W/V*).
2. Ammonium persulfate(1.6%, *W/V*) in H_2O.
3. Deionized H_2O.
4. Detergent, household dishwashing.
5. Ethanol.
6. KOIL/Methanol solution.
7. Silanizing fluid; Gel Slick (FMC Bioproducts), RainX (Unelko, Scottsdale Arizona), and Acrylease(Stratagene).
8. 10 × TBE electrophoresis buffer.
9. TBE.
10. *N*, *N*, *N'*, *N'* - tetramethylethylenediamine(TEMED).
11. Urea, solid.

Procedures

1. If necessary, remove old silanizing reagent from plates by swabbing them with KOH/methanol solution. To prevent contamination of the glass surfaces by skin oils, wear talc – free gloves at all times and handle the plates by their edges.

2. Wash the plates and spacers in a warm, dilute solution of dishwashing liquid and then rinse them thoroughly in tap water, followed by deionized H_2O. Rinse the plates with absolute ethanol to prevent water spots, and allow them time to dry.

3. Treat the inner surface of the smaller or notched plate with silanizing solution. Lay the plate, inner surface uppermost, on a pad of paper towels in a chemical fume hood, and pour or spray a small quantity of silanizing fluid onto the surface of the plate. Wipe the fluid over the entire surface with a pad of Kimwipes and then allow the fluid to dry in the air (1 – 2minutes). Rinse the plate first with deionized H_2O and then with ethanol, and allow the plate time to dry.

4. Lay the larger (or unnotched) glass plate (clean side up) on an empty test tube rack on the bench and arrange the spacers in place along each side of the glass plate so they are flush with the bottom of the plate.

5. Center the shorter (notched) plate, siliconized side down, on top of larger (unnotched) plate. Make sure that the spacers remain in position at the very edges of the two plates.

6. Clamp the plates together on one side with two or three large (5cm length) bulldog binder clips. Bind the entire length of the other side and the bottom of the plates with gel sealing tape to make a watertight seal.

7. Remove the bulldog clips and seal this side of the gel plates with gel – sealing tape.

8. Place the flat side of the sharkstooth comb into the open end of the gel mold so that it fits snugly. Remove the comb and lay the empty gel mold on the test – tube rack.

9. Cover the working area of the bench with plastic – backed protective paper.

10. In a 250ml side – arm flask, prepare a sequencing gel solution containing the desired concentration of acrylamide as specified in table 11 – 2. The volumes given in the table are sufficient for a single 40cm × 40cm sequencing gel and can be proportionally adjusted to accommodate smaller or larger gels. The preparation of the gel must be completed without interruption from this point onward.

Table 11 – 2 Acryl amide Solutions for Denaturing Gels

	4% GEL	6% GEL	8% GEL	10% GEL
Acrylamide : bissolution (45%)	8.9ml	13.3ml	17.8ml	22.2ml
10 × TBE buffer	10.0ml	10.0ml	10.0ml	10.0ml
H_2O	45.8ml	41.4ml	36.9ml	32.5ml
Urea	42.0g	42.0g	42.0g	42.0g

11. Combine all of the reagents and then heat the solution in a 55℃ water bath for 3minutes to help dissolution of the urea.

12. Remove the solution from the water bath and allow it to cool at room temperature for 15minutes.

Swirl the mixture from time to time.

13. Attach the side – arm flask to a vacuum line and de – gas the solution.

14. Transfer the solution to a 250ml glass beaker. Add 3. 3ml of freshly prepared 1. 6% ammonium persulfate and swirl the gel solution gently to mix the reagents.

15. Add 50 μl of TEMED to the gel solution and swirl the solution gently to mix the reagents. Pour the gel solution into the mold directly from the beaker in which it has been prepared. Alternatively, draw approx. 40ml of the solution into a 60ml hypodermic syringe. Do not suck air bubbles into the syringe! Work as quickly as possible from here onward because the gel solution will polymerize rapidly. Polymerization can be appreciably slowed by putting the gel solution on ice, a boon for inexperienced gel pourers.

16. Allow a thin stream of gel solution to flow from the beaker or syringe into the top corner of the gel mold while holding the mold at an approx. 45° to the horizontal.

17. Lay the mold down on the test – tube rack.

18. Immediately insert the flat side of a sharkstooth comb approx. 0. 5cm into the gel solution. Insert both ends of the comb into the fluid to an equal depth so that the flat surface is level when the gel is standing in a vertical position.

19. Clamp the comb into position using bulldog binder clips. Use the remaining acrylamide/urea solution in the hypodermic syringe/pipette to form a bead of acrylamide across the top of the gel. Allow the gel to polymerize for at least 15minutes at room temperature.

20. Wash out the 60ml syringe so that it does not become clogged with polymerized acrylamide.

21. After 15minutes of polymerization, examine the gel for the presence of a Schlieren line just underneath the flat surface of the comb. This is a sign that polymerization is occurring satisfactorily. When polymerization is complete (approx. 1hour after the gel was poured), remove the bulldog clips.

22. The gel can be used immediately or stored for up to 24hours at room temperature or 48hours at 4℃. To prevent dehydration during storage, leave the tape in place and surround the top of the gel with paper towels dampened with 1 × TBE. Cover the paper towels with Saran Wrap. Do not remove the comb at this stage.

Part D Loading and running DNA – sequencing gels

Make sure that the gel boxes used for electrophoresis are well insulated, that all buffer chambers are covered, and that the box is used on a stable bench top that is dry. Always turn off the power to the box before sample loading or dismantling the gel.

Buffers and solutions

1. 5 × TBE.
2. 1 × TBE.
3. DNA sequencing reactions.

Procedures

1. Use damp paper towels or a wet sponge to wipe away any dried polyacrylamide/urea from the

outside of the gel mold. Pipette several ml of 1 × TBE buffer along the top of the smaller or notched glass plate and slowly remove the comb from gel. Cut the electrical tape with a scalpel and strip it from the bottom of the gel mold. Do not remove the tape from the sides of the gel!

2. Attach the gel mold to the electrophoresis apparatus with bulldog binder clips, plastic – coated laboratory clamps, or, in the case of electrophoretic devices with built – in screw clamps, according to manufacturer's instructions. The smaller or notched plate should be in direct contact with the electrophoresis device. The larger, unnotched plate should face the investigator.

3. Fill the upper and lower reservoirs of the apparatus with the appropriate buffer.

(1) Fill the upper and lower buffer reservoirs with 1 × TBE. Make sure that the level of the buffer in the lower chamber is well above the bottom of the plates. The level of the buffer in the upper chamber should be well above the level of the upper edge of the shorter or notched plate and be in direct contact with the gel.

(2) Use a 10ml syringe filled with 1 × TBE to rinse the top of the gel. Make sure that excess polyacrylamide and urea are removed from the gel. If necessary, use a syringe needle to scrape off any polyacrylamide sticking to the glass plates. Remove air bubbles under the bottom of the glass plates in the lower reservoir using a Pasteur pipette.

(3) Attach the electrodes to the electrophoresis apparatus and the power supply. The cathode (black lead) should be attached to the upper reservoir and anode (red lead) to the bottom. Attach the built – in thermal sensor (if available) or temperature – monitoring strip. Run the gel at constant wattage (50 – 70W) for approx. 45minutes or until the temperature of the gel reaches 45 – 50℃. Turn off the power supply and disconnect the electrodes.

4. Incubate the microfuge tubes containing the sequencing reactions for 2minutes in a heating apparatus set at 100℃. If the reactions have been carried out in a microtiter plate, remove the cover from the plate and float the open plate in a water bath for 5minutes at 85℃.

5. While the tubes or plates are incubating, fill a 10ml syringe fitted with a 22 – gauge hypodermic needle with 0.5 × TBE or 1 × TBE, as appropriate. Squirt the TBE forcibly across the submerged loading surface of the gel to remove any remaining urea and fragments of polyacrylamide from the loading area. Continue squirting until no more urea can be seen in the loading area.

6. Gently insert the sharkstooth comb (teeth downward) into the loading slot. Push the comb down until the points of the teeth just penetrate the surface of the gel.

7. Transfer the microfuge tubes or microtiter plate from the water bath or heating block to ice. Keep the samples at 0℃ until they are loaded onto the gel. Quick – cooling to low temperature retards renaturation of the template and radiolabeled strands.

8. Load 1 – 5 μl of each sequencing reaction into adjacent slots of the gel.

9. When all of the samples have been loaded, connect the electrodes to the power pack and the electrophoresis apparatus, cathode (black) to the upper reservoir and anode (red) to the bottom reservoir. Run the gel at sufficient constant power to maintain a temperature of 45 – 50℃ (table 11 – 3).

Table 11 –3 The Relationship of Gel Size, Power and Voltage

Gel Size(cm)	Power(W)	Voltage(V)
20 ×40	35 –40	approx. 1700
40 ×40	55 –75	approx. 2000 –3000
40 ×40(electrolyte gradient gel)	75	approx. 2100

Monitor the progress of the electrophoretic run by following the migration of the marker dyes in the formamide gel –loading buffer(table 11 –4).

Table 11 –4 Migration Rates of Marker Dyes through Denaturing Polyacrylamide Gels

POLYACRYLA MIDE GEL(%)	BROMO PHENOL BLUE	XYLENE CYANOL FF
5	35	130
6	29	106
8	26	76
10	12	55
20	8	28

Typically, electrophoresis is continued until the xylene cyanol dye is within 5 –10cm of the bottom of the gel and the bromophenol blue dye is at the bottom. The numbers are the approximate sizes of DNA(in nucleotides) with which the indicated marker dye will comigrate in a standard DNA sequencing gel.

10. Depending on the distance between the sequence of interest and the oligonucleotide primer, apply a second loading of the sequencing samples to a standard or formamide –containing denaturing polyacrylamide gel approx. 15minutes after the bromophenol blue in the first set of samples has migrated to the bottom of the gel(1.5 –2.0hours). The sequence obtained from the first loading will be more distal to the primer, whereas that obtained from the second loading will be more proximal. The length of readable sequence can be extended by approx. 35% by reloading the samples into a fresh set of lanes 2hours after the samples were first loaded.

(1) Turn off the power supply and disconnect the sequencing apparatus.

(2) Replace the buffer in the top and bottom reservoirs.

(3) Denature the samples by heating as described in Step 4 above.

(4) Load the samples.

(5) Reconnect the sequencing apparatus to the power supply.

(6) Run the gel, as before, at sufficient constant power to maintain a temperature of 45 –50℃.

11. At the end of the run, follow the procedures described in Part E to dismantle the gel and perform autoradiography.

Part E Autoradiography and reading of sequencing gels

Buffers and solutions

1. Sequencing gel –fixing solution.

2. Radioactive ink. Reusable alternatives to radioactive ink are chemiluminescent markers available from Stratagene(Glogos). The markers can be used multiple times and should be exposed to fluorescent light just prior to a new round of autoradiography.

Procedures

1. At the end of the electrophoretic run, turn off the power and disconnect the sequencing apparatus from the power pack. Dispose of the electrophoresis buffer and then remove the gel mold from the apparatus.

2. Lay the gel mold flat on plastic - backed protective bench paper with the smaller(notched) plate uppermost. Allow the gel to cool to under 37℃ before proceeding.

3. Remove any remaining pieces of gel - sealing tape. Use the end of a metal spatula to pry apart the plates of the mold slowly and gently. The gel should remain attached to the longer(nonsiliconized) glass plate.

4. When the glass plates have been separated, cut off a bottom or top comer on the side of the gel that was loaded first. This landmark serves to orient the gel during subsequent manipulations.

5. Fix the gel in sequencing gel - fixing solution(methanol/acetic acid).

6. Transfer the gel together with its supporting glass plate, to a shallow tray containing methanol: acetic acid fixing solution. Fix the gel for 30minutes at room temperature. Do not agitate the fluid while the gel is being fixed.

7. After 30minutes, lift the glass plate very, very slowly from the fixation solution. Try to keep the plate horizontal until most of the fixation solution has drained away. Lay the plate, gel side uppermost, on a stack of paper towels. Blot excess fixation solution from the glass plate with Kimwipes. Try not to touch the surface of the gel with the Kimwipes. Remove wrinkles and blemishes from the gel by gently caressing its surface with gloved fingers.

8. Prepare a piece of Whatman 3MM CHR paper(or equivalent) that is slightly larger(2 - 3cm) than the gel in both length and width. Hold the paper in a bow shape and touch the center of the bow to the center of the gel. Let the paper fall gently onto the surface of the gel and then apply gentle pressure so that the gel becomes firmly attached to the rough surface of the paper.

9. Hold the paper in place with one hand and pick up the supporting glass plate with the other hand. Quickly flip the sandwich over and lay it down on the bench top with the glass plate facing upward. Gently separate the 3MM CHR paper from the glass plate by lifting the plate upward. The gel will stick to the 3MM CHR paper as the glass plate is pulled away.

10. Lay the 3MM CHR paper(gel uppermost) on two pieces of Whatman 3MM paper of the same size. Cut a piece of Saran Wrap slightly longer and wider than the gel and lay it on top of the gel. Try to avoid creases and bubbles. This step is more easily accomplished with the help of another person. Hold the corners of the Saran Wrap and pull outward so that it is tightly stretched. Lower the stretched Saran Wrap onto the surface of the gel. Once the Saran Wrap has touched the gel, do not attempt to remove it, since this can cause the gel to tear. The flat end of an agarose gel comb, a Kimwipe, or a plastic card can be used to remove any bubbles of air trapped between the gel and wrap.

11. Use a paper cutter or sharp pair of scissors to trim all three pieces of Whatman paper and

the Saran Wrap to approximately the same size as the gel.

12. Place the sandwich of paper, gel, and Saran Wrap on the gel dryer, with the plastic wrap uppermost. Use a sheet of Mylar to keep the sandwich flat during drying.

13. Following the instructions of the manufacturer of the gel dryer, dry the gel for 30 – 60minutes under vacuum at 80℃.

14. Remove the gel from the dryer and peel off the Saran Wrap. The dried gel should feel smooth to the touch and not sticky. A quick remedy to stickiness is to turn a powdered latex glove inside out and use the talcum powder on the inside of the glove to dust the gel. To orient the gel, attach a small adhesive label marked with radioactive or chemiluminescent ink to the 3MM CHR paper in the space created by cutting the bottom corner of the gel (Step 4).

15. In a darkroom, place the dried gel (gel side up) in a spring – loaded metal cassette. Cover the gel with a sheet of unexposed X – ray film. Close the cassette. Establish an autoradiograph by exposing the gel to the film for 16 – 24hours at room temperature or 80℃.

16. Develop the autoradiograph according to the recommendations of the manufacturer of the film and read the sequence of the DNA.

实验十二　RNA 分离与定量

扫码“学一学”

扫码“看一看”

【导言】

为了克隆某一基因、研究基因的调控、确定特定 mRNA 的大小与结构、鉴定基因表达产物，以及测定特定 mRNA 的合成与降解速率都需要分离特定的 RNA。选择 RNA 分离方法主要考虑以下因素：RNA 的来源、类型和相对丰度，样品的大小以及分离步骤是否简便易行。

提取的 RNA 的纯度及完整性对一些实验非常关键，如 RT－PCR，Northern 印迹，寡聚（dT）对 poly（A）$^{+}$RNA 的分离，cDNA 的合成以及体外翻译。这对于 cDNA 文库的构建尤为重要，因为 cDNA 文库的构建需要高纯度的 mRNA。

【目的】

通过本实验掌握 RNA 的分离与定量的方法。

【原理】

RNA 特别容易降解，因此在分离时要特别小心。在对完整 RNA 进行分离的所有方法中都使用强变性剂来抑制内源性核糖核酸酶（RNases）。与脱氧核糖核酸酶（DNases）不同，RNases 不易失活，因为它们不需要辅因子且热稳定性高，热变性后仍可重新折叠。

任何完整 RNA 的成功分离都需要以下 4 个重要步骤：细胞或组织的有效裂解；核蛋白复合物的变性；RNases 的灭活；从含有 DNA 和蛋白质的体系中纯化出 RNA。其中最关键的步骤是在细胞裂解时，必须立即将从细胞器膜上释放出的内源性 RNases 立即灭活。

本实验介绍如何使用 Trizol 来分离总 RNA。Trizol 是一种酸性苯酚提取试剂，含有去垢剂和使 RNases 失活的异硫氰酸胍。Trizol 法抽提很简单，它首先从细胞中抽提 RNA，然后再将其 DNA 分离。当细胞裂解并用苯酚抽提时，蛋白和脂类往往存在于酚相，而 RNA 存在于水相（与苯酚的 pH 无关）。相反，DNA 的分配则表现出 pH 依赖性，当苯酚的水相用缓冲液调至 pH 7.5 时，DNA 与 RNA 进入水相。然而，当苯酚中的水相用缓冲液调至 pH 4.0 时，DNA 则存在于酚层。RNA 在 260nm 有光吸收，因此很容易对 RNA 的浓度进行检测。A_{260}：A_{280}可用来表示 RNA 的纯度。纯净 RNA 的比值应为 1.8～2.1。然而即使比值低于 1.8，这种纯度的 RNA 也可使用，这可通过随后的凝胶电泳结果加以验证。

【试剂与器材】

1. 细胞（5×10^{6} HeLa）。

2. DEPC 水溶液：向 1L 去离子水中加 1ml DEPC，盖上塞子并振荡。静置数小时并高压灭菌。DEPC 有很强的毒性，应该密闭保存。

3. 氯仿（ACS grade）。

4. 异丙醇（ACS grade）。

5. 75% 乙醇（ACS grade 苯酚和 DEPC 水溶液）。

6. Trizol 溶液（一种酸性苯酚提取试剂，含有清洁剂和使 RNases 变性的异硫氰酸胍）。

【实验步骤】

（一）RNA 分离

1. 安全地除去细胞生长介质。

2. 小心地往表面皿加入 1ml PBS 来洗涤单层细胞，避免使细胞破裂。倒出 PBS 洗涤液。

3. 向表面皿中加入 1ml Tirzol 来裂解细胞。让细胞裂解物在移液管中上下移动数次（Tirzol 量不足会使 DNA 污染分离出的 RNA）。

4. 将溶解产物转移到一个 1.5ml 或 2.0ml 离心管中。

5. 将同源样本在室温下培育 5 分钟，使核蛋白复合物得到充分的分离。

6. 每 1ml Tirzol 中添加 0.2ml 氯仿，确保塞子塞紧。

7. 用手用力振摇离心管 15 秒并在室温下培育 2.5 分钟。

（二）RNA/DNA 分离

1. 在 4℃下低于 12 000rpm 离心 15 分钟。离心后，混合物分离，形成下层的苯酚 - 氯仿层（红色）、中间的分界面以及最上层的水层（无色）。RNA 单独存在于水层中。

2. 将水层转移至一个空试管中（上层约 600ml）。

3. 为使 RNA 沉淀，每 1ml Trizol 中加入 1ml 异丙醇用以初步裂解，倒置混合，然后在室温下培育 10 分钟。

4. 在 4℃下，以 12 000rpm 离心 10 分钟（冷冻室）。

5. 除去漂浮物并用 75% 乙醇（每 1ml Trizol 中至少加入 1ml 75% 乙醇，用以初步裂解）旋转洗涤 RNA 小球。

6. 在 4℃下，以 7500rpm 离心 5 分钟。

7. 除去漂浮物并通过翻转微量离心管 5 ~ 10 分钟风干 RNA 小球。避免使 RNA 完全干燥。

8. 将 RNA 溶解于大约 20μl 的无 RNase 的水中。

9. 55℃加热 RNA 5 ~ 10 分钟（提高 RNA 溶解度），-20℃ O/N 保存。

（三）RNA 纯度检测

1. 将 2 ~ 3μl RNA 溶于 500μl H_2O 中。

2. 置于石英的吸收池中，测量 OD_{260} 和 OD_{280}。

3. 用下列公式计算 RNA 浓度.

$$\text{RNA}(\mu g/\mu l) = A_{260} \times \frac{0.5}{\text{volume}(\mu l)\ \text{of RNA assayed}} \times 40$$

RNA 和 DNA 在 260nm 处有吸收峰。A_{260}：A_{280} 可用作 RNA 的纯度指示。纯净 RNA 的比值应为 1.8 ~ 2.1。然而当比值低于 1.8 时，这种纯度的 RNA 也可使用，这可通过凝胶电泳的结果得到证实。

Trizol 可从 GibcoBRL 中获得。如果不想使用苯酚，可购买 RNA 抽提试剂盒。

【注释】

1. 为了获得高质量的 RNA，需要创造一个无核糖核酸酶的环境。然而核糖核酸酶非常不易失活，因此 RNA 分离期间与之后都应特别小心，避免带入 RNases。如果最初的材料难以获得或不可替代，这一点则尤为重要。若要防止 RNases 污染样本以获得完整长度的 RNA，须注意以下几方面。

（1）在处理用于 RNA 分离分析的试剂时，应无菌操作。整个过程中都要戴手套，并要经常更换。

（2）在处理 RNA 时尽可能使用一次性无菌塑料制品。这些材料一般不含 RNases，并且无需进行使 RNases 失活的预处理。

（3）非一次性玻璃与塑料制品在使用前都应进行处理，以确保不带有 RNases。玻璃制品应在 200℃条件下烘烤过夜。塑料制品在使用前应用 0. lN NaOH/1mM EDTA 和稀释的焦碳酸二乙酯（DEPC）水溶液进行彻底冲洗，然后高温消毒 30 分钟以去除 DEPC。

DEPC 水溶液的制备：向 1L H_2O 中添加 1ml DEPC，混合并静置 0. 5 小时，过夜，高温消毒。DEPC 具有高度诱导突变的能力，必须密闭条件下处理。避免水流入 DEPC 储备瓶中，因为有水存在时 DEPC 会分解成为乙醇和 CO_2。CO_2压力的增加会导致爆炸，因此，尽量一次性用完准备的 DEPC。旧瓶子很危险。

（4）仅依靠高温消毒不足以使 RNases 失活。分离和电泳所使用的溶液应用 DEPC 进行处理，以去除 RNases。注意：DEPC 会与胺类反应，如 Tris，因此，Tris 及类似的缓冲液要用 DEPC 处理的水配制，而不是直接用 DEPC 处理。称量 Tris 时要小心，避免 RNases 污染，并在制备 Tris 缓冲液时使用 DEPC 处理过的水和玻璃器具。

（5）尽可能将 RNA 溶液冰冻保存，如果可能，将 RNA 以乙醇沉淀的小球形式保存于 -70℃超低温冰箱或 -70℃的甲酰胺中。甲酰胺的 pH 应大于 5. 2。

2. 异硫氰酸胍是很强的蛋白变性剂，但它不能使 RNA 酶发生不可逆的失活。因此，如果去蛋白后样本被残余的 RNA 酶污染，这些酶在变性剂去除后仍可重新恢复活性。将水相中的 RNA 彻底地从有机相中分离出，并避免通过污染容器或手套被蛋白重新污染，这一点很重要。

3. 异硫氰酸胍有剧毒，所以其操作应在通风橱中进行。避免接触皮肤，液体与固体废物要作为毒性废物处理。不要使用聚碳酸酯试管盛放异硫氰酸胍。

Experiment 12 RNA isolation and quantification

Introduction

RNA is specifically isolated in order to clone genes, study gene regulation, determine the size and structure of specific messages, identify gene products and determine the rates of synthesis and degradation of specific messages. The choice of an RNA isolation method depends upon a number of factors. The source of RNA, type of RNA to be purified, relative abundance of the RNA, sample size and convenience of the isolation procedure are all factors to consider.

The purity and integrity of isolated RNA is critical for its effective use in procedures such as RT – PCR, Northern blotting, oligo(dT) selection of poly(A) + RNA, cDNA synthesis, and translation*in vitro*. This is especially true for the construction of cDNA libraries where efficient cDNA synthesis requires highly purified mRNA.

Purpose

Upon completion of this exercise, you should be able to describe the process of RNA Isolation and Quantification.

Principle

RNA is notoriously susceptible to degradation and special care is required for its isolation. All methods of RNA isolation use strong denaturants to inhibit endogenous ribonuclease (RNases) for isolation of intact RNA. RNases, in contrast to deoxyribonu – cleases (DNases), are difficult to inactivate as they do not require cofactors and are heat – stable, refolding following heat denaturafion.

The successful isolation of intact RNA by any procedure requires that four important steps be performed: effective disruption of cells or tissue; denaturation of nucleoprotcin complexes; inactivation of RNases activity; purification of RNA away from contaminating DNA and protein. The most important of these is the immediate inactivation of endogenous RNase activity which is released from membrane bound organelles upon cell disruption.

In this experiment a method for isolation of total RNA using Trizol was introduced. Trizol (an acid phenol extraction reagent containing detergents and guanidine isothiocyanate to denature RNases) extraction is a simple procedure to extract RNA from cells, and to separate RNA from DNA. When cells are lysed and extracted with phenol, protein and lipids always partition into the phenol layer whereas RNA partitions to the aqueous layer (regardless of the pH of the phenol). In contrast, DNA shows a pH – dependent partitioning. When the aqueous phase of the phenol is buffered above pH 7.5, DNA partitions with RNA into the aqueous layer. However, when the aqueous phase of the phenol is buffered at pH 4.0, DNA will partition into the phenol layer. The concentration of RNA can be conveniently measured due to its ability to absorb light at 260nm. The ratio A_{260}: A_{280} is an indication of the purity of the RNA. The ratio (for pure RNA) should be 1.8 – 2.1. However, even if the ratio is less than 1.8, the RNA quality may be acceptable and can be confirmed by the subsequent gel electrophoresis.

Regents and apparatus

1. Cells (5×10^6 HeLa).

2. DEPC treated water: Add 1ml of DEPC to 1L of deionized water, cap, and shake. Let sit several hours and autoclave. DEPC is highly toxic and should be aliquoted in a hood.

3. Chloroform (ACS grade).

4. Isopropanol (ACS grade).

5. 75% Ethanol (ACS grade EtOH and DEPC treated water).

6. Trizol reagent (an acid phenol extraction reagent containing detergents and guanidine isothiocyanate to denature RNases).

Procedures

1. RNA isolation

(1) Discard the cell growth medium safely.

(2) Wash the cell monolayer by carefully by adding 1ml PBS to the side of the dish making sure you do not disrupt the cells. Pour off the PBS wash.

(3) Add 1ml Trizol to the plate to lyse the cells. Pass the cell lysate several times through a pipette (insufficient Trizol may result in contamination of the isolated RNA with DNA).

(4) Transfer the lysate to a 1.5ml or 2.0ml Eppendorf tube.

(5) Incubate the homogenized samples for 5min at room temperature. This allows complete dissociation of nucleoprotein complexes.

(6) Add 0.2ml of chloroform per 1ml of Trizol. Make sure the tube is tightly capped.

(7) Shake tubes vigorously by hand for 15 sec and incubate at room temp for 2.5min.

2. RNA/DNA separation

(1) Centrifuge samples at no more than 12 000rpm for 15min at 4℃. Following centrifugation the mixture separates into a lower, phenol - chloroform phase (red), an interface, and an aqueous upper phase (colourless) - RNA remains exclusively in the aqueous phase.

(2) Transfer the aqueous phase (i.e. top layer; about 600μl) to a fresh tube.

(3) To precipitate RNA, add 1ml isopropanol per 1ml of Trizol used for the initial lysis and mix by inversion, then incubate at room temp for 10min.

(4) Centrifuge at 12 000rpm for 10min at 4℃ (cold - room).

(5) Discard the supernatant and wash the RNA pellet with 75% ethanol (use at least 1ml of 75% ethanol per 1ml of Trizol used for the initial lysis) - mix by vortexing.

(6) Centrifuge at 7500rpm for 5min at 4℃.

(7) Discard the supernatant and (briefly) air - dry the RNA pellet by inverting the Eppendorf tube for 5 - 10min. It is important NOT to dry the RNA completely.

(8) Dissolve RNA in about 20ul RNase - free water.

(9) Heat RNA at 55℃ for 5 - 10min (increases solubility of RNA). Store at -20℃ O/N.

3. RNA concentration measurement

(1) Dilute 2 - 3μl of RNA (e.g. approx. 1/10 of RNA from above) into 500μl H_2O.

(2) Place in a quartz cuvette and measure the OD_{260} and OD_{280}.

(3) Calculate the RNA concentration using the following equation:

$$\text{RNA}(\mu g/\mu l) = A_{260} \times \frac{0.5}{\text{volume}(\mu l)\ \text{of RNA assayed}} \times 40$$

RNA and DNA has an absorbance peak at 260nm. $A_{260}: A_{280}$ is an indication of the purity of the RNA. The ratio (for pure RNA) should be 1.8 - 2.1. However, even if the ratio is less than 1.8, the RNA quality may be acceptable and can be confirmed by the subsequent gel electrophoresis.

Trizol can be purchased from GibcoBRL. If you wish to avoid phenol, many companies sell RNA extraction kits.

Notes

1. It is important to creat a Ribonuclease - Free environment for preparing high quality RNA.

However, Ribonucleases are extremely difficult to inactivate. Great care should be taken to avoid inadvertently introducing RNases into the RNA preparation during or after the isolation procedure. This is especially important if the starting material has been difficult to obtain or is irreplaceable. The following notes may be helpful in preventing the accidental contamination of the sample with RNases, allowing the isolation of full length RNA.

(1) To prevent contamination, sterile technique should be employed when handling any of the reagents used for RNA isolation or analysis. Wear gloves at all times and change them frequently.

(2) Whenever possible, use sterile, disposable plasticware for handling RNA. These materials are generally RNase – free and do not require pretreatment to inactivate RNases.

(3) Nondisposable glassware and plasticware should be treated before use to ensure that it is RNase – free. Glassware should be baked at 200℃ overnight. Plasticware should be thoroughly rinsed before use with 0.1N NaOH/1mM EDTA and with diethylpyrocarbonate (DEPC) – treated water, then autoclaved for 30minutes to remove any trace of DEPC.

DEPC treated water is made by adding 1ml DEPC per liter of H_2O. Mix and let sit 0.5h to overnight and autoclave. DEPC is highly mutagenic and should only be handled in a hood. Be careful not to allow water to enter the stock bottle of DEPC because in presence of water DEPC breaks down to ethanol and CO_2. The increase in CO_2 pressure can lead to explosions. For this reason, try to use up all the DEPC you order at one time. Old bottles are potentially hazardous!

(4) Autoclaving alone is not sufficient to inactivate RNases. Treat the solutions used for isolation and electrophoresis of RNA with DEPC to inactivate RNases. Note that DEPC reacts with amines such as Tris. Hence, Tris and similar buffers should be made from DEPC treated water, but not treated with DEPC directly. Use caution when weighing out Tris to avoid RNase contamination and use DEPC – treated water and glassware when preparing Tris buffers.

(5) Keep RNA solutions on ice on much as possible, and if possible store RNA as an ethanol – precipitated pellet at –70℃ or in formamide at –70℃. The formamide should have a pH >5.2.

2. Guanidine isothiocyanate is very powerful protein denaturant, but it does not irreversibly inactive RNases. Hence, if residual RNases contaminate the sample after deproteinization, these enzymes can become active again following the removal of the denaturant. Hence, it is important that the RNA containing aqueous phase be cleanly separated from the lower organic phase, and not be recontaminated by protein from dirty containers or gloves.

3. Guanidine isothiocyanate is VERY TOXIC. Therefore, manipulations with it should be carried out in a fume hood. Contact with skin should be avoided. Liquid and solid waste should be discarded as toxic waste. Do not use polycarbonate tubes with Guanidine isothiocyanate.

实验十三　Northern 印迹

扫码“学一学”

【导言】

Northern 印迹与 Southern 印迹杂交类似，但它用于分析 RNA。该技术主要用于鉴别在不同类型的细胞中哪些基因处于活跃状态。Northern 杂交首先要从细胞中分离出 RNA，其中包括该类细胞中所有活跃基因产生的信使 RNAs（mRNAs）。这些 mRNA 编码不同大小的蛋白，因而长短不一。细胞中编码高丰度蛋白的 mRNA 其含量也比编码低丰度蛋白的 mRNA 要多。

【目的】

通过本实验熟悉 Northern 印迹杂交的基本原理与杂交过程。

【原理】

RNA 可以通过琼脂糖/甲醛凝胶电泳按分子质量大小分开，泳动快的 RNA 片段较小，其迁移距离与 RNA 分子的大小成反比。RNA 电泳后经 EB 染色在紫外灯下可观察到。总 RNA 凝胶电泳后可见 28S 和 18S 核糖体亚基的条带，通常可作为分子质量标记（约 4. 8kb 和 1. 9kb）。

利用 Northern 杂交探测某一特定的 mRNA，首先需要将 RNA 从琼脂糖/甲醛凝胶转移到尼龙膜上。然后冲洗尼龙膜并将其用 X 线片进行曝光。胶片冲洗后通常能够显示出与探针识别相应的 mRNA 条带。利用条带相对于膜顶部的位置可估测出 mRNA 长度，条带灰度的深浅可指示出 mRNA 在原细胞中的丰度：表达量低的 mRNA 条带颜色浅，而高丰度 mRNA 条带颜色较深。因此，Northern 杂交为鉴别细胞中哪些基因被打开、表达水平以及 mRNA 大小提供了有用信息。

A 部分　RNA 的甲醛琼脂糖凝胶电泳

【试剂与器材】

1. 无 RNA 酶的水（如 DEPC 处理过的）。
2. 琼脂糖。
3. 10×缓冲液（0. 2M MOPS pH 7. 0，10mM EDTA）。
4. 甲醛。
5. 紫外灯。
6. 甲酰胺。
7. EB。

8. 0.5M 碘乙酰胺。
9. 微量离心管。

【实验步骤】

（一）样品准备

1. 将所有的样品置于冰上，向无菌的 Eppendorf 管中加入 10～20μg 来自对照细胞和（或）处理过的细胞的 RNA，并加入 DEPC 处理过的水至 15μl。

2. 向 Eppendorf 管中加入 2.5μl 去离子甲酰胺、2.5μl 甲醛和 2.5μl 电泳缓冲液。

3. 将样品于 60℃水浴 20 分钟做变性处理，然后立即置于冰上。

（二）凝胶准备

1. 将 130ml 琼脂糖（1.5g 琼脂糖/130ml 水）于 60℃熔化。

2. 小心加入 15.0ml 10×电泳缓冲液，3.0ml 0.5M 碘乙酰胺，2.25ml 40% 甲醛和 10.0μl 溴化乙锭（10mg/ml）。

3. 搅拌混匀，倒入封好的电泳槽中，注意要将梳子插入槽中。

（三）加样和电泳

将样品加入加样孔中，100V 电泳 2 小时。

B 部分　Northern 印迹

【试剂与器材】

1. Hybond N（尼龙）膜。
2. 20×SSC（0.3M 枸橼酸钠，3M NaCl）。
3. Saran 包装膜（DOW 化学公司）。
4. 玻板。
5. Whatman 3MM 滤纸。
6. 重物瓶。
7. 20×SSC 浸湿的海绵。

【实验步骤】

1. 停止电泳，通过 UV 发生器/EtBr 染色观察 RNA 的完整性。

（1）戴手套，从槽中移去凝胶，小心将凝胶置于 UV 透射仪中拍照。通常于照相前在凝胶边放一把荧光尺，有助于将照片上核糖体 Marker 的位置（也许不是实际大小）与最终 X 线片上杂交的 mRNA 进行比对。

（2）将凝胶从槽中转移到 UV 透射仪中。避免将电泳液溅出，因为其中可能含有 EB。

2. 将样品转移（O/N）到尼龙膜上。按图 13－1 所示建立转移装置，并通过毛细管作用转移过夜。

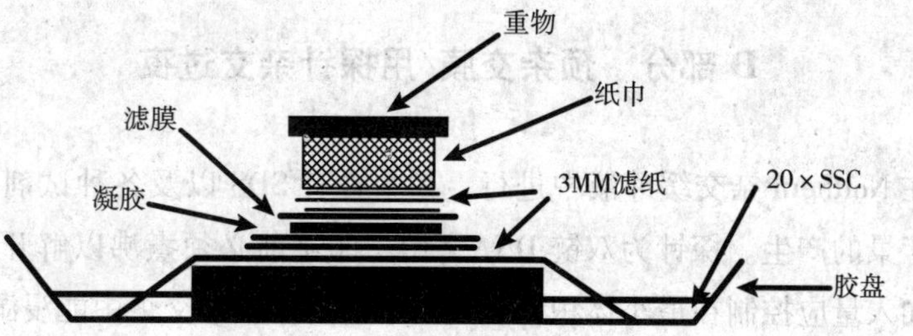

图 13－1　Northern 印迹装置建立

C 部分　探针的合成与纯化

Northern blot 可被任何标记的 RNA 或 DNA 探测到，此实验介绍用放射性标记的 DNA 进行探测。

【试剂与器材】

1. 2×SSC 缓冲液。
2. 50μCi ［α－^{32}P］dCTP，3000Ci/mM。
3. High Prime 溶液。
4. 带螺丝帽的小瓶。

【实验步骤】

（一）拆散转移装置

1. 仔细拆开转移装置，移去尼龙膜。
2. 利用紫外线交联的原理将 RNA 固定在尼龙膜上。
3. 标记杂交盒。
4. 将印迹膜置于杂交槽盒中（有 RNA 的一面向上）。

（二）探针合成

High Prime 溶液是一种含有随机寡聚核苷酸、Klenow 聚合酶、标记级的 dATP 和 dGTP 与 dTTP，以及在 50% 浓度甘油中可快速有效地用^{32}P 或^{35}S 标记的 dCTP 对 DNA 进行标记的优化的反应缓冲液。所用的质粒含有 P2X2 cDNA，可将该质粒作为模板与 High Prime 溶液以及［α－^{32}P］dCTP 一起使用。

1. 置 25ng 模板 DNA 于具盖螺口小瓶中（终体积 10μl）。
2. 沸水浴加热变性 DNA 10 分钟。
3. 冰上骤冷。
4. 于微量离心机中快速离心片刻。
5. 在冰上混合下列组分：25ng 变性 DNA；4μl High Prime 溶液；5μl 50μCi ［α－^{32}P］dCTP，3000Ci/mM。
6. 37℃保温 10 分钟。
7. 加 2μl 0.2M EDTA 终止反应，并于 65℃保温 10 分钟。

D部分　预杂交膜/用探针杂交过夜

印迹杂交在Northern杂交缓冲液中进行，其中含有SDS以及各种试剂，它们可以阻断各种非特异性背景的产生。探针为双链DNA，用于印迹前必须煮沸以解开双链。探针杂交时需过夜，其加入量应控制在最小体积，只要在水浴摇床或杂交炉中能浸湿杂交膜即可。

【试剂与器材】

1. 放射性标记探针。
2. 杂交缓冲液：50%甲酰胺，6×SSPE，5×Denhardt试剂，1% SDS。
3. （超）声波降解的鲑鱼精子DNA（10mg/ml）。

【实验步骤】

1. 将（超）声波降解的鲑鱼精子DNA煮沸5~10分钟，然后在冰浴中迅速冷却。加入杂交缓冲液中至终浓度为200mg/ml，如0.3ml鲑鱼精子DNA（10mg/ml）加入15ml杂交液。

2. 把杂交液（含加入的ssDNA）加入含有膜的杂交盒中。42℃预杂交至少1小时（无时间上限）。

3. 煮沸探针10分钟，冰浴中迅速冷却。

4. 打开杂交盒，在不搅动膜的情况下把杂交溶液倒入一个15~20ml的试管中（一般不充满15ml）。再将变性（煮沸过的）探针加入杂交缓冲液并温和混匀，然后把它加入杂交盒，并于42℃孵育过夜。

E部分　洗涤杂交膜并对胶片曝光过夜

【试剂与器材】

1. 2×SSC（含0.1% SDS）。
2. Saran包装膜。
3. X线胶片。
4. 0.1×SSC（含0.1% SDS）。

【实验步骤】

1. 将具放射性的杂交溶液倒入指定的容器（槽）中。膜留在杂交盒中并进行洗涤。

2. 向杂交盒中加入约50ml含有0.1% SDS的2×SSC（室温），并洗涤膜10分钟。

3. 重复上述洗涤步骤。

4. 倒掉第二次的洗涤液。向杂交炉中加入50ml预热到42℃含有0.1% SDS的2×SSC，洗涤20分钟。

5. 倒掉最后一次的洗涤液。将膜转移到干燥的Whatman滤纸上晾干，用Saran包装膜包裹后与 X 线胶片曝光过夜。24 小时后可冲洗胶片，如有必要可曝光更长时间。

【注意】

1. 甲醛有毒，是潜在的致癌物质，应在通风橱中使用以避免吸入。甲酰胺、溴化乙锭和碘乙酰胺也具有毒性，应戴手套处理。

2. 在没有护目镜保护的情况下，UV 射线会损伤眼。熔化的琼脂糖会导致烫伤，小心处理。

3. ^{32}P 标记的核苷酸会放射 β－粒子，应小心处理。将废弃液倒入特定的废物杯中。操作时戴两层手套。

4. Saran 包装膜可从 Sigma 公司购买。此类凝胶的标准操作中会用到高浓度的甲醛，不允许加入溴化乙锭，因此凝胶需要在后道工序中染色。本实验中使用的甲醛溶液浓度较低，可在胶中加入溴化乙锭。

5. 碘乙酰胺作为 RNA 酶抑制剂加入凝胶中较危险。

6. 膜上只有少量 RNA 包含 P2X2 mRNA，它可以与放射性探针杂交。但是膜本身带有强放射背景，需要进行洗涤才能得到干净的胶片和清晰的结果。

Experiment 13 Northern blotting

Introduction

Northern blotting is similar to Southern blotting but is used to analyze RNA rather than DNA. This technique is used to identify which genes are active in different cell types.

For Northern blotting, RNA is isolated from cells. This contains messenger RNAs (mRNAs) derived from all the active genes in that cell type. The mRNAs are different sizes, depending on the size of the protein they encode. mRNAs that encode proteins present at high concentrations in a cell are also more abundant than those encoding proteins present at low concentrations.

Purpose

Upon completion of this exercise, you should be familiar with the principle and process of Northern Blot.

Principle

The RNA is electrophoresed through an agarose/formaldehyde gel which separates the RNA species by size. The faster migrating RNA fragments are the smallest, however, the distance of migration is not linear, rather it is inversely proportional to the size of RNA molecule. When RNA has separated following electrophoresis, it is stained with ethidium bromide and visualised using ultra violet light. For gels of total RNA the 28S and 18S ribosomal subunits are visible and act as convenient size markers (approx. 4.8kb and 1.9kb, respectively).

To probe for a specific mRNA species by northern blot, it is first necessary to transfer the RNA from the agarose/formaldehyde gel to a nylon membrane. The membrane is then washed and exposed to X - ray film. Development of the film reveals usually a single band corresponding to the mRNA recognized by the probe. The position of the band relative to the top of the membrane can be used to estimate the length of the mRNA and the intensity or darkness of the band is a measure of how much of that mRNA was present in the original cells. Thus mRNAs of genes that are expressed at low levels appear as faint bands and very abundant mRNAs appear as dark bands. Northern blotting can, therefore, provide useful information about which genes are switched on in a cell, the level at which they are expressed and the sizes of the different mRNAs.

Part A Formaldehyde agarose gel electrophoresis of RNA

Reagents and apparatus

1. RNase free water(i. e. DEPC treated).
2. Agarose.
3. 10 × running buffer(0. 2M MOPS pH 7. 0, 10mM EDTA).
4. Formaldehyde.
5. UV transilluminator.
6. Formamide.
7. Ethidium bromide.
8. 0. 5M iodoacetamide.
9. Microfuge.

Procedures

1. Preparation of the samples

(1) Keeping all samples on ice, add 10 - 20μg of RNA from the control cells and(or) treated cells to a sterile Eppendorf tube. The volume of RNA should be increased to 15μl by the addition of DEPC treated water.

(2) To this add 2. 5μl deionized formamide, 2. 5μl formaldehyde and 2. 5μl of running buffer (10 × stock), and 2. 5μl loading buffer.

(3) Denature samples by heating at 60℃ for 20min, then snap cool on ice.

2. Gel Preparation

(1) You will be provided with 130ml molten agarose at 60℃ (1. 5g agarose/130ml H_2O).

(2) Carefully add: 15. 0ml 10 × running buffer, 3. 0ml 0. 5M iodoacetamide, 2. 25ml 40% formaldehyde, 10. 0μl Ethidium bromide(10mg/ml stock).

(3) Mix by swirling and pour into sealed gel tray. Don't forget the comb to form the wells.

3. Load and run the gel

Load samples into the wells and run at 100 volts for 2hours.

Part B Northern Blot

Reagents and apparatus

1. Hybond N(Nylon) membrane.
2. 20 × SSC(0.3M trisodium citrate, 3M NaCl).
3. Saran wrap(DOW chemical company).
4. Glass plate.
5. Whatman 3MM paper.
6. Bottle of water as weight for blot.
7. Sponge soaked in 20 × SSC.

Procedures

1. Stop gel and visualise integrity of RNA by UV illumination/EtBr staining

(1) Wearing gloves, remove gel from the tank, carefully place gel on transilluminator and take a photograph. It is often useful to place a fluorescent ruler on the gel prior to the photograph. This aids later in matching the position of the ribosomal markers on the photograph (which may not be actual size) with the final X - ray film of the hybridized mRNA.

(2) Transport gel from tank to UV transilluminator. Take care not to spill the running buffer because it may contain ethidium bromide (EtBr).

2. Transfer O/N to nylon membrane

Set up transfer as shown (Fig. 13 - 1) and leave to transfer by capillary action overnight.

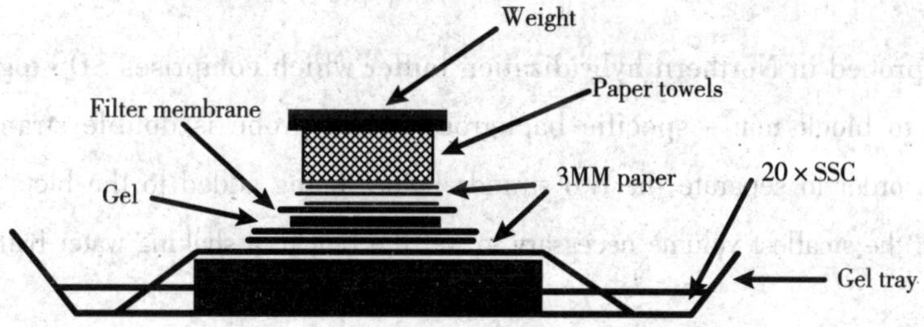

Fig. 13 - 1 Set - up for Northern Blot

Part C Synthesize and purify probe

The Northern blot can be probed with any labelled RNA or DNA. Here we will probe the blot with DNA which is radiolabelled.

Reagents and apparatus

1. 2 × SSC buffer.
2. 50μCi [α - ^{32}P] dCTP, 3000Ci/mM.
3. High Prime solution.

4. Screw cap vials.

Procedures

1. Transfer disassembly

(1) Carefully disassemble the transfer apparatus and remove the nylon membrane.

(2) Fix the RNA to nylon membranes using UV crosslinker.

(3) Label your hybridization chamber.

(4) Place blot in hybridization chamber (RNA "side" facing upwards).

2. Probe synthesis

High Prime is a reaction mixture that contains random oligonucleotides, Klenow polymerase, labelling grade dATP, dGTP, dTTP, and an optimized reaction buffer concentration in 50% glycerol for rapid and efficient labeling of DNA with ^{32}P – or ^{35}S – labeled dCTP. The plasmid we will use contains the P2X2 cDNA. We use this plasmid as the template DNA with High Prime and [α – ^{32}P] dCTP.

(1) Place 25ng template DNA (final volume 11μl) in a screw cap vial.

(2) Denature DNA by heating in boiling water bath for 10min.

(3) Chill quickly on ice.

(4) Pulse spin tube in microcentrifuge.

(5) Mix on ice: 25ng denatured DNA; 4μl High Prime solution; 5μl 50μCi [α – ^{32}P] dCTP, 3000Ci/mM.

(6) Incubate for 10min at 37℃.

(7) Stop the reaction by adding 2μl 0.2M EDTA and heat to 65℃ for 10min.

Part D Pre – hybridize membrane/hybridize with probe O/N

The blot is probed in Northern hybridization buffer which comprises SDS together with a variety of reagents to block non – specific background. The probe is double stranded DNA, so it must be boiled in order to separate the two strands before being added to the blot. The probe is applied overnight in the smallest volume necessary to wet the blot in a shaking water bath or hybridization oven.

Reagents and apparatus

1. Radiolabelled probe.

2. Hybridization buffer: 50% formamide, 6 × SSPE, 5 × Denhardt's reagent, 1% SDS.

3. Sonicated salmon sperm DNA (10mg/ml).

Procedures

1. Boil sonicated salmon sperm DNA for 5 – 10min and snap cool on ice. Add to hybridization buffer to a final concentration of 200μg/ml. e. g. 0.3ml of salmon sperm DNA (10mg/ml) to 15ml hybridization fluid.

2. Add the hybridization fluid (with added SS DNA) to the hybridization chamber containing the membrane. Prehybridize the membrane at 42℃ for at least 1h (no upper time limit).

3. Boil probe for 10mins and snap cool on ice.

4. Open the hybridization chamber, and without disturbing the membrane, pour the hybridization solution into a 15 – 20ml tube (you will not recover the full 15ml). Add denatured (boiled) probe to the hybridization buffer and mix gently, then return it to the hybridization chamber and incubate overnight at 42℃.

Fart E Wash membranes and expose to film O/N

Regents and apparatus

1. 2 × SSC (with 0.1% SDS).

2. Saran wrap.

3. X – ray film.

4. 0.1 × SSC (with 0.1% SDS).

Procedures

1. Discard the radioactive hybridization solution in the designated container/sink. Leave the membrane in the hybridization chamber for subsequent washes.

2. Add about 50ml 2 × SSC with 0.1% SDS (room temperature) to the hybridization chamber and wash the membrane for 10min.

3. Repeat the above washing step.

4. Discard the second wash solution. Add to the hybridization chamber, 50ml 0.1 × SSC with 0.1% SDS, preheated to 42℃. Wash for 20min.

5. Discard the final wash solution. Remove the membrane and air dry of a piece of Whatman filter paper. Wrap in Saran wrap and expose overnight to X – ray film. Gels may be developed after 24hours, or can be left for longer periods if necessary.

Notes

1. Formaldehyde is toxic and a potential carcinogen, use the fume hood – do not inhale! Formamide, ethidium bromide and iodoacetamide are also toxic and should be handled with gloves.

2. UV light can damage your eyes if not protected with facemask or goggles. Molten agarose can cause nasty bums, so handle with care.

3. β – particles are emitted by ^{32}P radionucleotides, need to be handled with care and disposed of in a designated radioactive waste bin. Wear two pairs of gloves.

4. Saran wrap can be purchased from Sigma. Standard protocols for these types of gels have high concentrations of formaldehyde that will not allow ethidium bromide addition so the gel has to be stained afterwards. In this course, we utilise a lower formaldehyde concentration which means that ethidium bromide can be included in the gel.

5. Iodoacetamide is included in the gel as an RNase inhibitor and is hazardous.

6. The small fraction of the RNA that comprises the P2X2 mRNA on the membrane should have hybridized to the radioactive probe. However, the membrane will have high background radioactivity which needs to be washed in order to yield a clean film and clear results.

扫码"学一学"

扫码"看一看"

实验十四 反转录－聚合酶链反应（RT－PCR）

【导言】

普通 PCR 使用热稳定性聚合酶且需要 DNA 模板，因而仅限于对 DNA 样品进行分析。然而很多情况下需要对来源于有机体中的 RNA 进行扩增，尤其是要对进化过程中不同组织中基因的表达差异进行分析以及对稀有 mRNA 来源的 cDNA 进行克隆时，RT－PCR 可间接地对某一特定的 RNA 分子进行扩增。

【目的】

通过本实验掌握 RT－PCR 的工作原理与操作方法。

【原理】

反转录 PCR 过程分成相对独立的两个步骤：将 mRNA 反转录成 cDNA（RT）和通过聚合酶链反应扩增 cDNA（PCR）。反转录包括由总 RNA 或 poly（A）$^+$RNA（mRNA）合成 cDNA，因此 cDNA 就为热稳定性聚合酶提供了必要的 DNA 模板，常用禽类成髓细胞瘤病毒（AMV）反转录酶或鼠类白血病病毒 Moloney 株（M－MLV 或 MuLV）反转录酶从 RNA 反转录合成 cDNA，用特定的寡核苷酸作引物就可以扩增出 cDNA 中的某个片段。

起始 RNA 模板的质量和纯度对于 RT－PCR 能否成功至关重要，总 RNA 和 poly(A)$^+$RNA 都可以作为起始模板，但它们都必须完整，而且未被任何基因组 DNA 污染。特定 poly（A）$^+$RNA 的捕获实际上意味着富集了 mRNA 靶信息，因而只需要较少的反转录反应就可以满足随后的扩增反应。第一条链的合成效率与 RNA 模板质量有关，又很大程度上对随后的扩增结果产生影响。

【试剂与器材】

1. 12～18 个碱基长度的寡聚 dT（100μg/ml）溶于 pH 8.0 的 TE 中。
2. 随机六核苷酸（1mg/ml）溶于 pH 8.0 的 TE 中。
3. 模板 RNA（100μg/ml 的总 RNA 水溶液或 10μg/ml 的 PolyA$^+$RNA 水溶液）。
4. 特定引物（300mmol/L）。
5. 10×扩增缓冲液。
6. 含 4 种脱氧核苷酸的 dNTP（20mmol/L，pH 8.0）。
7. 胎盘 RNA 酶抑制剂（20 单位/微升）。
8. 外源参比 RNA。
9. PCR 仪。
10. 微量离心管。
11. 0.5ml PCR 薄壁管。

【实验步骤】

（一）反转录（RT）

1. 取一个无菌 500μl Eppendorf 管按以下顺序加样：

总 RNA（0.5μg）	X
dNTPs（10mmol/L）	1μl
12～18bp 寡聚 dT（0.5μg/ml）	1μl
用 DEPC 处理的水	加至 10μl

2. 65℃下恒温 5 分钟。

3. 缓慢冷却至室温。

4. 另取一只微量离心管按以下顺序加样：

10×RT－PCR 缓冲液	2μl
$MgCl_2$（25mmol/L）	4μl
0.1mol/L DTT	2μl
RNA 酶抵制剂	1μl

5. 将此混合物加入第一步中的 RNA－dNTP－寡 dT 混合物中。

6. 42℃恒温 2 分钟。

7. 加 1μl（50 单位）Superscript Ⅱ聚合酶。

8. 42℃恒温 50 分钟。

9. 70℃恒温 15 分钟（使 Superscript Ⅱ聚合酶失活），然后冰水冷却。

10. 脉冲自旋。

11. 加入 1μl RNA 酶抑制剂，并于 37℃恒温 20 分钟。

（二）聚合酶链反应（PCR）

1. 取一只 PCR 薄壁管按以下顺序加样：

25mmol/L $MgCl_2$	3μl
10×RT－PCR 缓冲剂（不含 $MgCl_2$）	5μl
反转录产物（cDNA）	2μl
5′引物或正向引物（300mmol/L）	1μl
3′引物或反向引物（300mmol/L）	1μl
dNTP 混合物（最终浓度 800μmol/L）	1μl
Taq 聚合酶（1 单位）	0.4μl
水	至最终体积 50μl

2. 将样品加入预先设置好程序的 PCR 仪器，总共运转约 80 分钟，具体参数如下。

（1）第一次变性　94℃，3 分钟；1 次循环。

（2）扩增循环　①变性：94℃，30 秒。②退火：60℃，30 秒。③延伸：72℃，30 秒。45 秒循环。

（3）最后延伸　72℃，5 分钟；1 次循环。

3. 从反应管和对照管中分别取 5～10μl 样品，进行琼脂糖凝胶电泳或聚丙烯酰胺凝胶电泳对其分析，其中包括一定大小的 DNA 标准对照，溴化乙锭或 SYBR Gold 染色后观察

DNA，如果反应成功应可观察到与预期大小一致的 DNA 片段。

Experiment 14 Reverse transcription – polymerase chain reaction (RT – PCR)

Introduction

The thermostable polymerases used in the basic PCR process require a DNA template and, as such, the technique is limited to the analysis of DNA samples. Yet numerous instances exist in which the amplification of RNA from an organism would be preferred. This is particularly true in analyses involving the differential expression of genes in tissues during development or the cloning of cDNAs from rare messages.

Purpose

Upon completion of this exercise you should be able to describe how RT – PCR works.

Principle

RT – PCR is used to indirectly amplify a specific sequence from RNA molecules. We will break the RT – PCR reaction into two discrete steps, RT (reverse transcription to convert mRNA to cDNA) and PCR (to amplify cDNA by the polymerase chain reaction). The RT reaction involves synthesizing cDNA from total RNA or polyA enriched RNA (mRNA), and this cDNA provide the necessary DNA template for the thermostable polymerase. Avian myeloblastosis virus (AMV) or Moloney murine leukemia virus (M – MLV or MuLV) reverse transcriptases are generally used to produce a DNA copy of the RNA template. Using specific oligonucleotides (i. e. primers) a fragment of the cDNA can be amplified.

The quality and purity of the starting RNA template is crucial to the success of RTPCR. Either total RNA or poly(A)$^+$RNA can be used as the starting template, but both must be intact and free of contaminating genomic DNA. Specific capture of poly(A)$^+$RNA will enrich a targeted message so that less of the reverse transcription reaction is needed for the subsequent amplification. The efficiency of the first strand synthesis reaction, which can be related to the quality of the RNA template, will also significantly impact the results of the subsequent amplification.

Reagents and apparatus

1. Oligo(dT)12 – 18 (100μg/ml) in TE (pH 8.0).

2. Random hexanucleotides (1mg/ml) in TE (pH 8.0).

3. Template RNA [The template may be total RNA (100μg/ml) in H_2O or poly(A)$^+$ RNA (10μg/ml) in H_2O].

4. The gene – specific primers (300nM each).

5. 10 × Amplification buffer.

6. dNTP solution(20mM) containing all four dNTPs(pH 8.0).

7. Placental RNase inhibitor(20units/μl).

8. Exogenous reference RNA.

9. PCR thermocycler.

10. Microfuge.

11. 0.5ml thin walled PCR tubes.

Procedures

1. Reverse transcription(RT)

(1) To a sterile 500μl Eppendorf tube add in the following order:

Total RNA(0.5μg)	Xμl
dNTPs(10mmol/L)	1μl
Oligo(dT) 12-18(0.5μg/ml)	1μl
DEPC treated H_2O	to 10μl

(2) Incubate 65℃ for 5min.

(3) Cool slowly to room temperature.

(4) Prepare the following in a separate tube:

10×RT-PCR buffer	2μl
$MgCl_2$(25mmol/L)	4μl
0.1mmol/L DTT	2μl
RNase inhibitor	1μl

(5) Add this mixture to the RNA/dNTP/oligo mixture from step 1.

(6) Incubate at 42℃ for 2min.

(7) Add 1μl(50units) Superscript Ⅱ polymerase.

(8) Incubate at 42℃ for 50min.

(9) Heat at 70℃ for 15min(to inactivate Superscript Ⅱ polymerase), then chill on ice.

(10) Pulse-spin.

(11) Add 1μl RNaseH and incubate at 37℃ for 20min.

2. Polymerase chain reaction(PCR)

(1) To a sterile 0.5ml thin walled PCR tube add in the following order:

25mmol/L $MgCl_2$	3μl
10×RT-PCR buffer(minus $MgCl_2$)	5μl
RT reaction(cDNA)	2μl
5′ or forward primer(300mmol/L)	1μl
3′ or forward primer(300mmol/L)	1μl
dNTP mix(final concentration 800μmol/L)	1μl
Taq polymerase(1unit)	0.4μl
H_2O	to a final volume of 50μl

(2) Place your sample into a pre-programmed thermal cycler, the running time is approximately 80min and 30 sec as described below.

1) First denature: 94℃ for 3min, 1cycle.

2) programme amplification cycle: ①94℃ for 30sec. ②60℃ for 30sec. ③72℃ for 30sec. 45 sec cycles.

3) End with an extension: 72℃ for 5min, 1cycle.

(3) Withdraw a sample (5 - 10μl) from the test reaction mixture and the control reactions and analyze them by electrophoresis through an agarose or polyacrylamide gel. Include DNA markers of an appropriate size. Stain the gel with ethidium bromide or SYBR Gold to visualize the DNA. A successful amplification reaction should yield a readily visible DNA fragment of the expected size.

实验十五　克隆化基因在大肠埃希菌中的表达

扫码“学一学”

扫码“看一看”

【导言】

基因工程和生物技术的一个主要目标是生产生物活性蛋白。它是通过在不同的宿主中对已克隆的基因进行表达来实现的。大肠埃希菌是最早用来生产重组 DNA（rDNA）药物的宿主系统，1982 年由 Eli Lilly 公司上市的重组人胰岛素就是在该系统中生产的。至今大肠埃希菌仍然是生产重组蛋白的一种重要宿主系统，并在工业生产中广泛应用。它具有多种优点，如易培养、低成本、高蛋白产量等。

【目的】

通过本实验帮助学生理解大肠埃希菌中克隆基因表达的基本原理。

【原理】

重组蛋白表达需要使用特定的表达载体，它含有能指导克隆化基因在宿主细胞中转录和表达的 DNA 信号序列。其他一些载体序列则可影响表达蛋白的稳定性或将表达蛋白分泌到胞外。为实现高水平表达，所用质粒载体必须在宿主细胞中以多拷贝形式存在。不同蛋白高效表达所需的条件也不相同，必须个别地进行优化，而这又导致产生了多种表达系统。大肠埃希菌能使多种蛋白得到快速而低廉的表达，故应用广泛。细菌表达系统仍然十分重要，许多已商业化的蛋白就是利用该系统进行生产的。

细菌表达载体具有许多重要特征，具体如下。①位于克隆序列上游的启动子能启始转录。选择一个与 RNA 聚合酶高度亲和的启动子对确保高表达水平至关重要。②位于克隆基因下游的转录终止序列。③位于转录起始位点下游的核糖体结合位点（SD 序列），核糖体与该序列结合的强弱很大程度上影响蛋白合成水平的高低。④克隆 DNA 序列的首位应有一个 AUG 翻译起始密码子，末位应有一个终止密码子。插入序列应该包含一个开放阅读框架。建立一个正确的阅读框架很重要，它决定能否翻译出正确的氨基酸序列。一般载体可使克隆化序列以不同的方式进行克隆，以便 3 种不同的阅读框架中的任何一种都可使用。⑤所有载体都应允许克隆序列进行诱导表达。连续的表达会导致细胞能源的枯竭，进而损伤细胞的其他功能。如果表达在某些特定时间被开启，而在其他时间被关闭，其表达效率会更高。大多数表达系统是通过向其培养基中添加一种小分子物质，通常为代谢中间产物，进行诱导。还有些系统则是通过改变培养温度进行诱导。

【试剂与器材】

1. IPTG——可诱导型表达载体。
2. 可转化和可携带 $lacI^{q}$ 或 $lacI^{q1}$ 等位基因的大肠埃希菌宿主菌。
3. 含重组表达质粒的大肠埃希菌工程菌。

4. 考马斯亮蓝染色剂和银染色剂。
5. IPTG（1M）。
6. 1×SDS 凝胶上样缓冲液。
7. 目的基因或目的 cDNA 片段。
8. 含 50μg/ml 的氨苄西林的 LB 琼脂平板。
9. 含 50μg/ml 的氨苄西林的 LB 液体培养基。

【实验步骤】

1. 挑取重组克隆至含适当抗生素的 50ml LB 液体培养基中，37℃下培养过夜。

2. 第二天再移取 1～2ml 培养物至 50ml 新鲜的含抗生素的 LB 液体培养基中，37℃下培养至 A_{600} 达 0.6。

3. 为了测定含可表达质粒的细胞的百分比，用对数生长期的细菌做质粒稳定性实验。

（1）分别将 1000μl、147μl 和 135μl 的 LB 培养基加入 3 支 1.5ml 无菌的微量离心管中。转移 1ml 对数生长期培养液至含 1000μl 的 LB 液体培养基的试管中，做初次稀释（稀释液 B）。取 3μl、15μl 初次稀释液，分别加入含 147μl（稀释液 A）、135μl（稀释液 B）的 LB 液体培养基中，做第二次稀释。

（2）按表 15－1 所示将细菌铺平板。

（3）37℃下将平板培养过夜。

4. 加 100mM IPTG 到对数生长期的细菌培养液中，至终浓度 0.4mM。

5. 保温培养，并在诱导过程中的不同时间段（如 1 小时、2 小时、4 小时和 6 小时）各取 1ml 培养液至微量离心管中，用分光光度计测其 A_{550}；室温下用微量离心管以最大转速离心 1 分钟，吸弃上清。

6. 加入 100ml 的 1×SDS 凝胶加样缓冲液，重悬沉淀；100℃加热 3 分钟；室温下最大转速离心 1 分钟；置于冰上，直至收集完全部样品，上样，做 SDS 凝胶电泳，分析重组蛋白的表达水平。

表 15－1 质粒稳定性测试实验

细菌体积	LB 平板中的附加成分	说明
50μl 的稀释液 A	不添加	所有的细胞都应形成菌落
50μl 的稀释液 A	含抗生素	>90% 的细胞形成菌落
50μl 的稀释液 B	含 IPTG（0.4mM）	<2% 的细胞形成菌落
50μl 的稀释液 B	含 IPTG（0.4mM）+抗生素	<0.01% 的细胞形成菌落

Experiment 15 Expression of cloned genes in *E. coli*

Introduction

One of the primary aims of genetic engineering and biotechnology is to produce biologically useful proteins. This is achieved by expressing cloned genes in a variety of host cells. *E. coli* was the first host used to produce a recombinant DNA (rDNA) pharmaceutical, enabling the approval of Eli Lilly's rDNA human insulin in 1982. Until now it is still an important host organism for the production of recombinant proteins and widely used in industrial production. Its many advantages include easy cultivation, low cost and high production potential etc.

Purpose

The objective of this exercise is to help the students understand the basic principle of the expression of the cloned genes in *E. coli*.

Principle

The expression of recombinant proteins requires the use of specialized vectors called expression vectors which contain DNA signal sequences that direct the transcription and translation of the cloned sequence by the host cell. Other vector sequences are included which influence the stability of the expressed proteins or direct the host cell to secrete the expressed protein. To ensure high levels of expression, plasmid vectors are used that exist in host cells as multiple copies. The conditions required for efficient expression vary from protein to protein and must be optimized individually. This has led to the development of many diverse expression systems. The bacterium *E. coli* allows many proteins to be produced rapidly and inexpensively and has been used extensively. Bacterial expression systems remain important and are used for the production of most commercially important proteins.

Vectors for expression of proteins in bacteria have a number of essential features: ①A promoter placed upstream of the cloned sequence to initiate transcription. Selection of a promoter that binds RNA polymerase strongly is necessary to ensure high levels of expression. ②A transcription termination signal placed downstream of the cloned sequence. ③A ribosome binding site (Shine - Dalgarno sequence) placed downstream of the transcription start site. The level of protein synthesis is influenced by how strongly the ribosome binds this sequence. ④The cloned DNA sequence should have an AUG translation initiation codon at the start and a termination codon at the end. The intervening sequence should be an open reading frame. It is very important to establish the correct reading frame to ensure that the correct amino acid sequence is translated. Vectors usually allow sequences to be cloned in different ways so that each of the three reading frames can be used. ⑤Vectors should allow expression of the cloned sequence to be induced. Continuous expression creates a drain on the energy resources of the cell which is detrimental to other cell functions. Expression is more efficient if it can be switched on for a limited period then switched off again. Most systems are induced by the addition of a small molecule, often a metabolite, to the bacterial culture. Some systems are induced by altering the culture temperature.

Reagents and apparatus

1. IPTG——inducible expression vector.
2. *E. coli* strain suitable for transformation and carrying either the *lacI*^{q} or *lacI*^{q1} allele.
3. Engineered *E. coli* strain harboring the recombinant expressin plasmid.
4. Coomassie Brilliant Blue stain or Silver stain.

5. IPTG(1 M).

6. 1 × SDS gel – loading buffer.

7. Gene or cDNA fragment of interest.

8. LB agar plates containing 50μg/ml ampicillin.

9. LB medium containing 50μg/ml ampicillin.

Procedures

1. Pick recombinant colonies and grow them overnight at 37℃ in 50ml of LB medium containing the appropriate antibiotic.

2. The next day remove 1 – 2ml and inoculate into 50ml of fresh LB medium containing antibiotic. Allow the culture to grow at 37℃ to an A_{600} of 0.6.

3. To determine the percentage of cells that contain expressible plasmid, perform the following Plasmid Stability Test on the log phase culture.

(1) Aliquot 1000μl, 147μl and 135μl of LB medium into three sterile 1.5ml microcentrifuge tubes. Prepare a primary dilution of bacteria by transferring 1ml of the log phase culture into the tube containing 1000μl of LB medium. Prepare two secondary dilutions of the bacteria by transferring 3μl and 15μl of the primary dilution into the tubes containing 147μl (dilution A) and 135μl (dilution B) of LB medium, respectively.

(2) Plate the bacteria as indicated in table 15 – 1.

Table 15 – 1 Plasmid Stability Test

Volume of Bacteria	LB Pate Additives	Comments
50μl of dilution A	No additives	Almost all cells should form colony
50μl of dilution A	Antibiotic	Expect > 90% of the cell to form a colony
50μl of dilution B	IPTG(0.4mM)	Expect < 2% of the cell to form a colony
50μl of dilution B	Antibiotic + IPTG(0.4mM)	Expect < 0.01% of the cell to form a colony

(3) Incubate the plates overnight at 37℃.

4. Add 100mM IPTG to the log phase bacterial culture to a final concentration of 0.4mM.

5. Incubate the culture and at various time points during the induction period (e.g. 1hour, 2hour, 4hour and 6hours), transfer 1ml of each culture to a microfuge tube, measure the A_{550} in a spectrophotometer, and centrifuge the tubes at maximum speed for 1minute at room temperature in a microfuge. Remove the supernatants by aspiration.

6. Resuspend each pellet in 100μl of 1 × SDS gel – loading buffer, and heat the samples to 100℃ for 3minutes. Centrifuge the tubes at maximum speed for 1minute at room temperature in a microfuge, and store them on ice until all of the samples are collected and ready to load on a SDS – gel to evaluate the expression level of the recombinant protein.

实验十六 SDS－聚丙烯酰胺凝胶电泳

扫码“学一学”

扫码“看一看”

【导言】

许多分子生物学研究都需在某一研究阶段纯化、鉴定和分析蛋白质。十二烷基硫酸钠－聚丙烯酰胺凝胶电泳（SDS－PAGE）则是其中应用最为普遍的技术。

【目的】

本实验在于帮助学生了解 SDS－聚丙烯酰胺凝胶电泳的基本原理及其应用。

【原理】

本技术中，待分离的样品在有 SDS 和还原剂的情况下通过加热变性并被去污剂所包被。SDS 可使蛋白质带上大量的负电荷，其带电荷的多少与多肽链的长度成正比。聚丙烯酰胺凝胶上样后，在高电压的作用下，蛋白组分向正极（阳极）迁移。

由于所有蛋白所带的负电荷与其分子大小成正比，蛋白就仅根据其分子质量的大小不同而被分离，即凝胶的分子筛效应。蛋白质的分子质量就可以通过与标准蛋白的凝胶迁移率进行比对而得到。理想的蛋白分离条带可通过不连续凝胶系统得到，这种不连续系统包括浓缩胶和分离胶，它们具有不同的盐浓度或 pH，或两者皆不相同。

凝胶电泳后，特定的凝胶条带可以通过染色而观察到，并且可通过被动扩散或电洗脱从凝胶上进行回收。两种最常用的染色方法是考马斯亮蓝法和银染法，这两种方法的蛋白条带检测灵敏度可分别达到 50ng 和 5ng。由于考马斯亮蓝法使用简便，并可以对收集的蛋白带做进一步分析，较为常用。考马斯亮蓝法和银染法都需要数小时才能完成。

【试剂与器材】

1. 50% 丙烯酰胺/双丙烯酰胺（29：1）用水溶解（48.3g 丙烯酰胺、1.7g 双丙烯酰胺制成 100ml，储存在深色瓶中）。注意：丙烯酰胺是一种强力可积累的神经毒素，制备丙烯酰胺时戴手套和穿防护服。

2. 分离用的凝胶缓冲液（lM Tris－HCl，pH 8.8）。

3. 10% SDS。

4. 浓缩用的凝胶缓冲液（0.375M Tris－HCl，pII 6.8）。

5. 10% 过硫酸铵。

6. 12% 分离胶配制方法（25ml，足够供 4 份 1.0mm 的微型凝胶）：50% 丙烯酰胺/双丙烯酰胺（29：1）溶解成 6.0ml；分离缓冲液（1M Tris－HCl，pH 8.8）9.4ml；10% SDS 250μl；水 9ml；TEMED 20μl；10% 过硫酸铵 300μl。

7. 4% 浓缩胶配制方法（12.5ml，足够供 4 份 1.0mm 的微型凝胶）：50% 丙烯酰胺/双（29：1）溶解成 1.0ml；浓缩缓冲液（0.375M Tris－HCl，pH 6.8）4.2ml；10% SDS

125μl；水 6.7ml；TEMED 20μl；10% 500μl。

8. 电泳缓冲液（10×）：29.0g Tris；144.0g 甘氨酸；10.0g SDS。用水定容到 1000ml。无须调节 pH。

9. 加样缓冲液（2×）：2.5ml 0.5M Tris－HCl，pH 6.8；2.0ml glycerol；4.0ml 10% SDS；0.5ml 0.1% 溴酚蓝；0.5ml β－巯基乙醇，用水定容到 10ml。

10. 低毒染色溶液：0.25g 考马斯亮蓝 R－250；100ml 乙醇；100ml 水，搅拌至溶解，加 25ml 醋酸，用水配成 250ml。

【实验步骤】

A 部分

1. 用胶带粘好凝胶槽底部，如果需要，竖立在架子上。

2. 在小锥形瓶中准备 12% 的溶解好的凝胶混合物。

3. 按顺序加 TEMED 和过硫酸铵，搅拌至彻底混合均匀。

4. 将混合物吸入带有细套管和针头的注射器中，注入凝胶槽中直至离槽顶部不足 2cm。在离心管中加 1～2ml 剩余的凝胶混合物，盖紧盖子。

5. 操作迅速，用注射器加 2～3ml 蒸馏水于每个槽的凝胶顶部。

6. 让凝胶聚合 1 小时。等待期间可以把离心管中的凝胶混合物作为参照估算槽中的凝胶何时聚合完全。

7. 在小锥形瓶中制备 4% 的浓缩胶混合物。按顺序加入 TEMED 和过硫酸铵，充分搅拌混匀。

8. 将混合物吸入带有细套管和针头的注射器中，注入凝胶槽中直至槽顶部。在离心管中加 1～2ml 剩余的凝胶混合物，盖紧盖子。

9. 将梳子插入每个凝胶槽中，注意避免在梳子下方留有气泡。

10. 聚合 1 小时。等待期间可以把离心管中的凝胶混合物作为参照估算槽中的凝胶何时聚合完全。

B 部分　制备聚丙烯酰胺凝胶电泳样品

为了制取澄清的样品，首先将等体积的蛋白缓冲溶液和 2 倍 SDS－PAGE 上样缓冲液混合，上样前加热至 95℃ 5 分钟，对于整个细胞（如 *E. coli*），取 1ml 过夜培养的菌体至微量离心管中，加 25μl 水和 25μl 2×SDS－PAGE 上样缓冲液。上样前加热至 95℃，14 000×g 离心 3 分钟去除细胞碎片。

C 部分　电泳以及凝胶观察

1. 用 10×SDS－PAGE 电泳缓冲液稀释制备 800ml 1×SDS－PAGE 电泳缓冲液。

2. 从凝胶槽中移去梳子，用 1×SDS－PAGE 电泳缓冲液并彻底冲洗梳孔以除去未聚合的丙烯酰胺。

3. 移去凝胶槽底部的胶带，将其放入凝胶电泳装置中。

4. 在中间腔室中加入 1×SDS－PAGE 电泳缓冲液并没过凝胶梳孔的顶部，检查是否会泄漏。

5. 上样 20μl 到梳孔中，根据需要，可使用微量注射器或带有凝胶进样头的进样器进样。

6. 将剩余的 1×SDS－PAGE 电泳缓冲液加满外腔室，贴近凝胶装置的顶部。

7. 连接电源，在 250V 直流电流下进行电泳，电泳应在 40～45 分钟内完成（可通过指示染料迁移至凝胶底部来指示）。

8. 凝胶电泳完成后，关闭电源并断开电源接线。

9. 从凝胶仪中取出凝胶槽，用小刀将凝胶从槽中分离出来，切去凝胶底部不规则的凝胶（底部缝隙中形成）和浓缩胶。

10. 将凝胶转移到染色缸中，并用 200～250ml 的染色液将其完全浸没进行染色。

11. 稍微遮盖后于微波炉中微波 2 分钟，再用振荡器缓慢振荡 10～15 分钟。此外，如果需要凝胶可于室温下染色过夜。

12. 染色完成后，倾去染液（染液可以循环使用）并用蒸馏水漂洗凝胶和容器以除去多余的染色液。

13. 加 200～250ml 脱色液，置于振荡器中缓慢振荡直至可以看见条带。这一过程一般需要 30～60 分钟。

Experiment 16　SDS－polyacrylamide gel electrophoresis

Introduction

Most molecular biology researches need to purify, identify or analyze proteins. At some research stage. SDS－PAGE (sodium dodecyl sulfate polyacrylamide gel electrophoresis) is the most commonly used technique for such purpose.

Purpose

The objective of this exercise is to help the students understand the basic principle of SDS－polyacrylamide gel electrophoresisthe and its application.

Principle

In this technique, the sample to be fractionated is denatured and coated with detergent by heating in the presence of SDS and a reducing agent. The SDS coating gives the protein a high net negative charge that is proportional to the length of the polypeptide chain. The sample is loaded on a polyacrylamide gel and high voltage is applied, causing the protein components to migrate toward the positive electrode (anode).

Since all of the proteins have a net negative charge that is in proportion to their size, the proteins are separated solely on the basis of their molecular mass——a result of the sieving effect of the gel matrix. The molecular mass of a protein can be estimated by comparing the gel mobilities of a band with protein standards. Sharp banding of the protein components is achieved by using a discontinuous gel system, having stacking and separating gel layers that differ in either salt concentration or pH or both.

Following gel electrophoresis, individual gel bands may be visualized and recovered from the gel by passive diffusion or electroelution. The two most commonly used visualization methods are Coomassie Blue and silver staining, which allow detection of protein bands to sensitivities of 50ng and 5ng, respectively. Coomassie Blue is more often used due to its simplicity of use and because recovery of the protein bands for further analysis is possible. Both Coomassie Blue and silver staining methods require several hours to complete.

Reagents and apparatus

1. 50% Acrylamide/Bis(29 : 1) Solution (48.3g acrylamide; 1.7g bis - acrylamide. Make to 100ml with water. Store in a dark bottle). Caution: wear gloves and protective clothing when preparing acrylamide gels! Acrylamide is a potent, cumulative neurotoxin.

2. Separating Gel Buffer (1M Tris - HCl, pH 8.8).

3. 10% SDS.

4. Stacking Gel Buffer (0.375M Tris - HCl, pH 6.8).

5. 10% Ammonium Persulfate.

6. Recipe for 12% separating gel (25ml, enough for four 1.0mm minigels): 50% Acrylamide/Bis (29 : 1) Solution 6.0ml; Separating Gel Buffer (1 M Tris - HCl, pH 8.8) 9.4ml; 10% SDS 250μl; water 9.0ml; TEMED 20μl; 10% Ammonium Persulfate 300μl.

7. Recipe for 4% stacking gel (12.5ml, enough for four 1.0mm minigels): 50% Acrylamide/Bis (29 : 1) Solution 1.0ml; Stacking Gel Buffer (0.375 M Tris - HCl, pH 6.8) 4.2ml; 10% SDS 125μl; water 6.7ml; TEMED 20μl; 10% Ammonium Persulfate 500μl.

8. Running Buffer (10 ×): 29.0 g Tris; 144.0g glycine; 10.0 g SDS. Make to 1000ml with water. Do not adjust pH.

9. Loading Buffer (2 ×): 2.5ml 0.5M Tris - HCl, pH 6.8; 2.0ml glycerol; 4.0ml 10% SDS; 0.5ml 0.1% bromophenol blue; 0.5ml β - mercaptoethanol. Make to 10ml with water.

10. Low Toxicity Staining Solution: 0.25g Coomassie Blue R - 250; 100ml ethanol; 100ml water. Stir until dye is dissolved. Add 25ml acetic acid and make to 250ml with water.

Procedures

PART A

1. Tape bottom of gel cassettes, if necessary, and arrange upright in a rack.
2. Prepare a 12% resolving gel mixture in a small Erlenmeyer flask.
3. Add the TEMED and ammonium persulfate in that order, and mix thoroughly by swirling.
4. Take up the mixture in a syringe, attach a narrow cannula or needle, and fill the gel cassettes

with the gel mixture to within 2cm of the top. Place 1 – 2ml of remaining gel mixture in a microcentrifuge tube and cap tightly.

5. Working quickly, gently layer 2 – 3mm of distilled water on the top of the gel mixture in each cassette, using a syringe and needle.

6. Allow the gel to polymerize for about 1 hour. You may use the gel mixture in the microcentrifuge tube as a guide as to when polymerization is complete.

7. Prepare a 4% stacking gel mixture in a small Erlenmeyer flask. Add the TEMED and ammonium persulfate in that order, and mix thoroughly by swirling.

8. Take up the mixture in a syringe, attach a narrow cannula or needle, and fill the gel cassettes with the gel mixture all the way to the top. Place 1 – 2ml of remaining gel mixture in a microcentrifuge tube and cap tightly.

9. Insert well – forming combs into each gel cassette, taking care not to trap bubbles under thecomb.

10. Allow the gel to polymerize for about 1 hour. You may use the gel mixture in the microcentrifuge tube as a guide as to when polymerization is complete.

PART B Preparing samples for SDS – polyacrylamide gel electrophoresis

For clarified liquid samples, mix equal volumes of protein – containing solution and 2 × SDSPAGE loading buffer. Heat sample to 95℃ for 5min prior to loading. For whole cells (*E. coli*), pellet 1ml of overnight culture in a microcentrifuge tube, and add 25μl of water and 25μl of 2 × SDS – PAGE loading buffer. Heat sample to 95℃ prior to loading, and centrifuge at 14 000 × g for 3min to pellet cell debris before loading.

PART C Running and visualizing gels

1. Prepare 800ml of 1 × SDS – PAGE running buffer from 10 × SDS – PAGE running buffer.

2. Remove the comb from a previously prepared gel cassette, and rinse the wells thoroughly with 1 × running buffer to remove any unpolymerized acrylamide.

3. Remove the tape from the bottom of the gel cassette, and place it into the gel appraratus.

4. Fill the middle chamber with 1 × running buffer above the top of the wells in the gel. Check for leaks into the outer compartment.

5. Load up to 20μl of sample into wells, as desired, using either a microsyringe or a micropipettor equipped with a gel – loading tip.

6. Fill the outer compartment with the remainder of the 1 × running buffer, and affix the top of the gel apparatus.

7. Attach the leads to a power supply and run gel(s) at a constant 250V. The gel run should be complete (as indicated by tracking dye migration to the bottom of the gel) in about 40 – 45min.

8. When the gel run is completed, switch off the power supply and disconnect the power leads.

9. Remove the gel cassette from the apparatus, and using a knife, separate the halves of the gel

cassette. Cut off the foot of the gel (which protrudes into the slot at the bottom) and the stacking gel.

10. Place the remaining portion of the gel in a container and complete cover with about 200 – 250ml of staining solution.

11. Cover loosely, and microwave at full power for 2min. Place on a shaker and slowly agitate for 10 – 15min. Alternatively, the gel can be safely stained overnight at room temperature, if desired.

12. When staining is complete, pour off the stain (it can be recycled and used many times) and rinse the gel and its container with distilled water to remove excess stain solution.

13. Add 200 – 250ml of destaining solution. Place on a shaker and slowly agitate until destaining is adequate to visualize bands. Typically only 30 – 60min is required.

实验十七　Western 印迹

扫码“学一学”

扫码“看一看”

【导言】

Western 印迹是从复杂的混合组分中对特定的蛋白质进行鉴定并确定其分子质量的一种技术。其名称的由来是这样的：DNA 印迹法最初是由 E. M. Southern 建立的，因此称为 Southern 印迹法，RNA 印迹法则被称为 Northern 印迹法。因此，蛋白质印迹法就理所当然地被称为 Western 印迹法了。

【目的】

通过本实验掌握 Western 印迹的原理以及操作过程。

【原理】

混合物经凝胶电泳分离后，分离的条带可转移到多聚膜上（印迹），然后将可快速检测的抗体加入，如果存在待阐明的蛋白质，抗体就会与其结合，这样就可以对其进行鉴别。此过程可以分为以下几个步骤。

1. 根据分子质量大小通过聚丙烯酰胺凝胶电泳对混合物中的变性蛋白进行分离。
2. 将分离的蛋白质转移到硝酸纤维膜（或尼龙膜）上，并保持其相对位置。
3. 通过和放射性标记或酶标记的抗体或配体的特定反应来检测蛋白质，并通过与已知分子质量的标准蛋白质加以比较来确定其分子大小。

实际操作中可能还有一些中间步骤。有些操作者习惯于对原始的凝胶进行染色以确证蛋白质分离效果满意。如果这样，就必须小心选择随后的操作方法，因为通常使用的蛋白质染色方法大多会使蛋白质发生沉淀，这样就会影响蛋白质有效地转移到膜上。然而，大多数有经验的实验操作者省略染色这一步，因为电泳本身往往可信。一种确证蛋白质已被凝胶有效分离的方法是使用一种商品化的彩色的分子质量标记，它可适时追踪分离效果而无需进行任何染色。在任何情况下，大多数操作者要对膜上的总蛋白质进行染色，从而监控蛋白质的转移效率。这对于保证所观察到的结果不会被特定区域蛋白质的转移失败影响过大非常重要。

【试剂与器材】

1. 细胞裂解物或其他来源的待测蛋白。
2. Immobilon P 膜（Millipore 公司 IPVH304 – FO）。
3. 第一抗体。
4. 与辣根过氧化物酶（HRP）偶联的第二抗体。
5. SuperSignal™ CL – HRP 底物系统（Pierce）。
6. Hyperfilm™ – ECL 胶片（Amersham）。

7. 氨基黑染料溶液（Sigma 公司 A－8181）。

8. 氨基黑脱色缓冲液：10% 甲醇；10% 醋酸。

9. 封闭缓冲液：5% Carnation 脱脂干奶粉（TBST 配制）。

10. TBS/Tween，TBST：25mM Tris－HCl，pH 8.0；125mM NaCl；0.1% Tween 20。

11. 抗体结合缓冲液：1% Carnation 脱脂干奶粉（TBST 配制）。

12. 微型凝胶电泳系统和转移系统，胶片洗片机。

【实验步骤】

1. 准备样品/对照品，并用 SDS－PAGE 进行分离。

2. 将蛋白质从凝胶转移到 Immobilon P 膜上，操作步骤依据转移系统生产厂家的说明书进行，以期获得最佳的转印结果。

3. 用氨基黑染色转移膜 1 分钟，然后将膜转移到装有氨基黑脱色缓冲液的盘中振荡 15～20 分钟。确认蛋白质已转移完毕。注释：这一步是为了确保蛋白质已从凝胶有效地转移到膜上。氨基黑可以与膜上所有的蛋白结合，并且不会影响随后的抗体与特定蛋白之间的反应。脱色后，寻找已完全转移的蛋白标记和实验泳道；或转移后用考马斯亮蓝对凝胶进行染色以确保膜的转移已经完成。

4. 转印膜脱色后，立即置于封闭缓冲液中振荡 1 小时。

5. 将转印膜与第一抗体在抗体结合缓冲液中混合后 4℃ 存放过夜。按照供应商提供的方法制备第一抗体，或使用 1～2μg/ml 的标准浓度。如果是测试杂交瘤组织培养物的上清液，需要将上清液与封闭缓冲液按 1∶5 的比例稀释。

6. 膜和第一抗体过夜处理后，用 TBST 冲洗 5 次，每次 5 分钟。

7. 在室温下将转印膜与偶联到 HRP 上的第二抗体作用 1 小时。按照抗体结合缓冲液中供应商推荐的方法制备第二抗体。

8. 与第二抗体作用后，在 TBST 中冲洗转印膜 5 次，每次洗 5 分钟。

9. 按照 Pierce 的 Chemiluminescence HRP 底物说明对转印膜进行冲印。

10. 将转印膜在 X 线片中曝光适当时间。为了得到最佳结果，对于特定的抗原抗体反应，可曝光 10 秒、1 分钟、5 分钟和 20 分钟来观察化学发光信号。

Experiment 17 Western blotting

Introduction

Western blotting is a technique for identifying a specific protein in a complex mixture and simultaneously determining its molecular weight. The name came about this way: DNA blotting was first described by E. M. Southern and, hence, became Southern blotting. RNA blotting then was dubbed Northern blotting. Thus, it was inevitable that protein blotting would become Western blotting.

Purpose

Upon completion of this exercise, you should be able to describe the principle and process of

Western Blotting.

Principle

When the mixture is separated by gel electrophoresis, and the separated bands are transferred (by "blotting") onto a polymer membrane. A readily detectable antibody is then added and if the protein in question is present, this will bind to it, so identifying it. The procedure can be broken down into a series of steps.

1. Size separation of the denatured proteins in the mixture by Polyacrylamide Gel Electrophoresis(PAGE).

2. Transfer of the separated proteins onto a nitrocellulose(or nylon) membrane while retaining their relative position.

3. Detecting the protein under investigation by its specific reaction with a radioactively labeled or enzyme labeled specific antibody or ligand and determination of its size relative to standard proteins of known size.

In practice there are usually a few intermediate steps. Some workers stain the original gel to ensure that the protein separation was satisfactory. If this is the case, the method must be chosen carefully because most of commonly used protein staining methods precipitate the proteins, and hence preclude efficient transfer of the proteins to a membrane. Most experienced workers, however, usually omit this step because it is usually quite reliable. One way to confirm that the proteins have separated properly on the gel is to use one of the commercially available colored molecular weight markers, which allow you to follow the separation in real time without any staining at all. In any case, most workers do monitor the efficiency of the protein transfer by staining the membrane for total protein. This is important to ensure that the results seen are not unduly influenced by the failure of proteins in particular area of the gel from transferring.

Reagents and apparatus

1. Cell lysates or other source of protein to be tested.
2. Immobilon P membranes(Millipore cat. no. IPVH304 – FO).
3. Primary antibody.
4. Secondary antibody conjugated to HorseRadish Peroxidase(HRP).
5. SuperSignal™ CL – HRP Substrate System(Pierce).
6. Hyperfilm™ – ECL films(Amersham).
7. Amido Black Staining Solution(Sigma cat. no. A – 8181).
8. Amido Black destaining buffer: 10% Methanol; 10% acetic acid.
9. Blocking Buffer: 5% Carnation nonfat dry milk in TBST.
10. TBS/Tween, TBST: 25mM Tris – HCl, pH 8.0; 125mM NaCl; 0.1% Tween 20.
11. Antibody Binding Buffer: 1% Carnation nonfat dry milk in TBST.
12. A mini – gel electrophoresis system and a transfer system and a film processor.

Procedures

1. Prepare samples/controls and separate proteins using an SDS – PAGE gel.
2. Transfer proteins from the gel onto Immobilon P membranes following instructions provided

by the transfer system manufacturer for best protein transfer results.

3. Stain the blot with Amido Black for one minute. Then, transfer the membrane to a dish containing the Amido Black Destaining Buffer and rock for 15 – 20minutes. Examine the blot for the complete protein transfer. Note: This step is incorporated to ascertain efficient protein transfer from the gel to the membrane. Amido Black binds to all proteins on the membrane and does not interfere with the subsequent antibody reaction with the specific protein. After destaining, look for the complete transfer of your protein marker and your test lanes. Alternatively, after the transfer, the gel can be stained with Coomassie blue stain to make sure that the transfer to the membrane has been completed.

4. After destaining the blot, immediately place it into the blocking buffer and rock for one hour.

5. Incubate the blot with primary antibody in the Antibody Binding Buffer overnight at 4 degrees C. Prepare the primary antibody as recommended by the supplier or use a standard concentration of 1 – 2μg/ml. If testing hybridoma tissue culture supernatant, dilute the supernatant 1 : 5 with blocking buffer.

6. After the overnight incubation of the membrane with the primary antibody, wash the blot five times in TBST, each wash for 5minutes.

7. Incubate the blot with the secondary antibody conjugated to HRP for one hour at room temperature. Prepare the secondary antibody as recommended by the supplier in the Antibody Binding Buffer.

8. After incubation of the secondary antibody, wash the blot 5times in TBST, each wash for 5minutes.

9. Develop the blot following the Pierce Chemiluminescence HRP substrate instruction.

10. Expose the blot to X – ray film for the appropriate time period. For best results, expose for ten seconds, one minute, 5minutes and 20minutes to visualize the chemiluminescence signal corresponding to the specific antibody – antigen reaction.

实验十八　计算机网络技术在分子生物学中的应用

扫码“学一学”

【导言】

互联网极大地改变了科研人员搜索及交换信息的方式。实验数据以前只能印在纸上进行交流，现在却可以通过电了的形式从集中的数据库中获得；期刊也以电子版的形式在网上出版；随着基因组、后基因组研究的深入，海量生物信息的收集、存储和分析已经离不开计算机和网络技术。

当今互联网上有数以亿计的网站，如何才能找到有用的数据？如何去判断哪些信息是有用的？通过本实验将学习如何使用互联网，从中获得生物科学研究所需要的数据。

【目的】

1. 掌握检索核酸和蛋白质序列的方法。
2. 掌握检索蛋白质结构信息的方法，并了解蛋白质结构的可视化。

【原理】

（一）数据库及序列数据检索

EMBL，GenBank 和 DDBJ 是国际上三大主要的核酸序列数据库。EMBL 由欧洲分子生物学实验室（European Molecular Biology Laboratory）于 1982 年创建。GenBank 由美国国家健康研究院（National Insiturte of Health，NIH）于 20 世纪 80 年代初建立，后移交国家生物技术信息中心（National Center for Biotechnology Information，NCBI）。DDBJ 是 DNA Data Base of Japan 的简称，创建于 1986 年，由日本国家遗传学研究所负责管理。1988 年，EMBL，GenBank 与 DDBJ 共同成立了国际核酸序列联合数据库中心，建立了合作关系，这三个数据中心各自搜集测序机构和实验室发布的序列数据，并通过计算机网络每天将新发现或更新过的数据进行交换，以保证这三个数据库序列信息的完整性。

本实验以 NCBI 中数据库的检索为例，介绍核酸和蛋白质序列的检索方法（表 18－1）。

表 18－1　NCBI 中主要的数据库

数据库	介绍
Nucleotide	包含来自 GenBank，EMBL 和 DDBJ 三大国际序列中心的核酸序列
Protein	包括从 GenBank，EMBL 和 DDBJ 核酸序列推测出的蛋白质序列，以及直接提交 Protein Information Resource（PIR），SWISSPROT，Protein Research Foundation（PRF）和 Protein Data Bank（PDB）（已解出结构的序列）等数据库中的蛋白质序列
Genome	提供不同基因组、完整染色体数据，以及片断重叠群序列图谱和整合的基因和物理图谱
Structure 和 MMDB	包含从结晶学和 NMR 实验得到的结构数据。分子建模数据库（Molecular Modeling DataBase，MMDB）的数据来自 Protein Data Bank（PDB）

续表

数据库	介绍
PopSet	包含来自某一种群、系统发生或突变研究得到的并经比对的成套序列数据。比对结果揭示了进化或种群间的差异，包含核酸和蛋白质序列数据
OMIM	在线人类孟德尔遗传数据库（Online Mendelian Inheritance in Man，OMIM）为人类基因和遗传疾病目录
Taxonomy	包含在 NCBI 遗传数据库中至少有一条核酸或蛋白质序列的所有生物名称
ProbeSet	是 NCBI 基因表达数据库（Gene Expression Omnibus，GEO）的 Entrez 系统版本。GEO 收集了基因表达和杂交列阵相关数据
3D Domains	囊括 NCBI 保守结构域数据库中的蛋白质结构域
UniSTS	是非冗余序列标签库
EntrezSNP	收集单核苷酸取代或短的删除或插入多态性
UniGene	是一个实验系统，将 GenBank 序列自动划分到确定方向的基因簇中，每一个 UniGene 含有某独一无二的基因序列，同时包括相关的基因被表达的组织型和基因图谱等信息
PubMed	PubMed 中心（PMC）提供电子版《生命科学杂志》论文全文

在检索时只要在下拉菜单中选择拟检索的数据库，然后输入关键词就可进行检索。多数搜索引擎使用布尔逻辑查询表单。布尔检索法把检索词用操作符 AND，OR 和 NOT 连接，将数据库返回的检索结果进行限制。这些操作符的含义如其字面意思：将两个关键词用 AND 连接，表明查找同时含有这两个关键词的文档；用 OR 连接，表明含两个关键词中间任何一个（或全部）的文档；而 NOT 表明寻找含关键词 1，但不含关键词 2 的文档。布尔检索法与文本相同，是从左到右阅读，可用括号来构成复杂的检索条件。例如，希望检索出的文档中有关键词 1，含关键词 2 或关键词 3 中的一个，但是没有关键词 4，输入的检索条件为：[关键词 1 AND（关键词 2 OR 关键词 3）] NOT 关键词 4。还可以使用引号来限定一个词组，如想要检索 protein sequence 在一起顺序出现的文档，可用“protein sequence”来限制检索结果。

（二）蛋白质结构检索及结构浏览器

早在序列数据库诞生之前的 20 世纪 70 年代，蛋白质结构数据库（Protein Data Bank，PDB）就已经问世。PDB 数据库原来由美国 Brookhaven 国家实验室负责维护和管理。1998 年，由美国国家科学基金委员会、能源部和卫生研究院资助，成立了结构生物学合作研究协会（Research Collaboratory for Structural Bioinformatics，RCSB），PDB 数据库改由 RCSB 管理。PDB 数据库以文本文件的方式存放数据，每个分子各用一个独立的文件。除了原子坐标外，还包括物种来源、化合物名称、结构提交者以及有关文献等注释信息。此外，还给出分辨率、结构因子，温度系数、蛋白质主链数目、配体分子式、金属离子、二级结构信息、二硫键位置等与结构有关的数据。PDB 数据库以文本文件格式存放，可以用文字编辑软件查看，但用文字编辑软件查看注释信息不太方便，更无法直观地了解分子的空间结构。

在过去，生物大分子成像是不可能的。1972 年 Levinthal 和 Katz 开创性地建立了计算机线框（wireframe）模型。此后，人们利用计算机程序实现了大分子立体结构的可视化，并可以拖动旋转等，为计算机辅助药物设计提供了帮助。常用的结构浏览器有 VRML，RasMol 和 Chime 等。此处仅介绍 Swiss－PdbViewer，它是一个免费的对蛋白进行显示和分析的软件，提供了大量的菜单以满足查看蛋白结构的需求，可以对活性部位及相关结构进行比较。

该软件可以在 Swiss Institute of Bioinformatics（SIB）主页上下载（网址为 https：//spdbv. vital－it. ch/）。

（三）使用 BLAST

Basic Local Alignment Search Tool（BLAST）是一套用于从所有序列数据库（包括核酸和蛋白质）中搜索相似性的程序。BLAST 程序兼顾检索速度和远源序列关系灵敏性，其给出的分值有良好的统计学解释，能使确实相似的序列从随机的背景中被选取出来，帮助推断输入序列可能的功能。BLAST 有几种比较方法可选择，见表 18－2。

表 18－2　BLAST 程序

程序	用途
BLASTP	氨基酸序列与蛋白数据库中序列比较
BLASTN	核酸序列与核酸数据库中序列比较
BLASTX	一个核酸序列不同读框翻译产物与一个蛋白数据库中蛋白序列比较，可以用来发现未知核酸可能的蛋白产物
TBLASTN	一个多肽序列与核酸数据库序列所有读框方式翻译成的蛋白序列比较。
TBLASTX	一个核酸序列的六种读框与一个核酸数据库的六种读框比较，但由于计算太复杂在网页中不能应用

【器材】

PC 最低配置介绍如下。CPU：酷睿 i3/内存 4GB；操作系统：win32 或更高版本，预装了 Internet Explorer，并接入互联网。

【实验步骤】

（一）检索 Pubmed

在浏览器地址栏输入 NCBI 的网址 https：//www. ncbi. nlm. nih. gov/，其首页界面如图 18－1 所示。以查找与水蛭素相关的文献为例，首先在下拉子菜单中选择 Pubmed，然后在

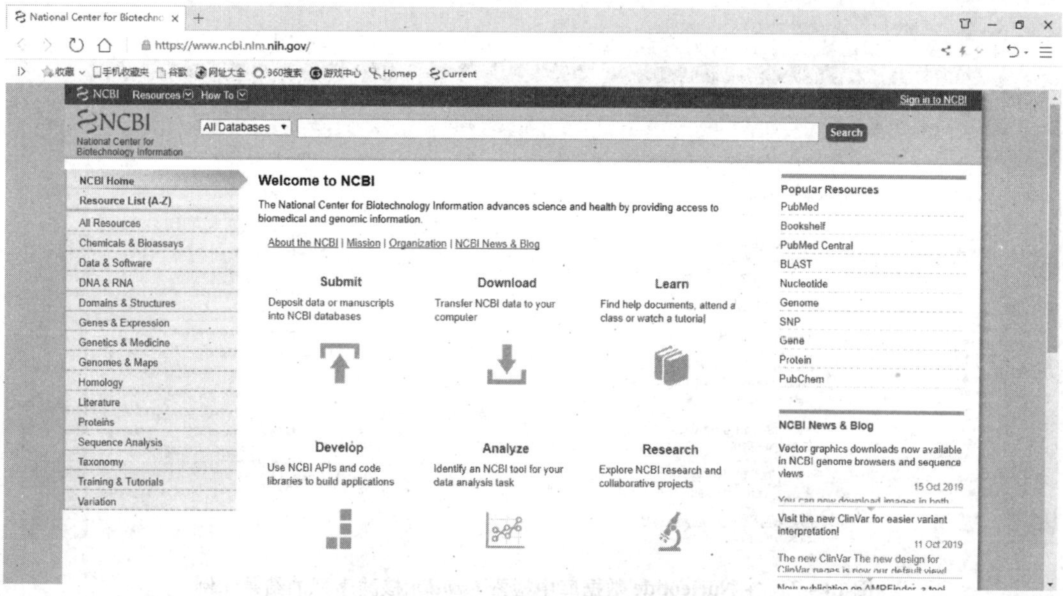

图 18－1　NCBI 的主页面

检索对话框中输入关键词 *hirudin*，点击 *search* 按钮或键盘上 *Enter* 键，结果如图 18－2 所示，表明本次检索得到 4705 条与水蛭素相关的记录，本页显示 20 项。由于检出条目较多，可用布尔表达式缩小检索范围，如检索水蛭素相关基因的克隆和表达，可用 *hirudin* AND *clone* AND *expression* 为检索词。在检索时可尝试不同的关键词，从而使检索结果更贴近需求。

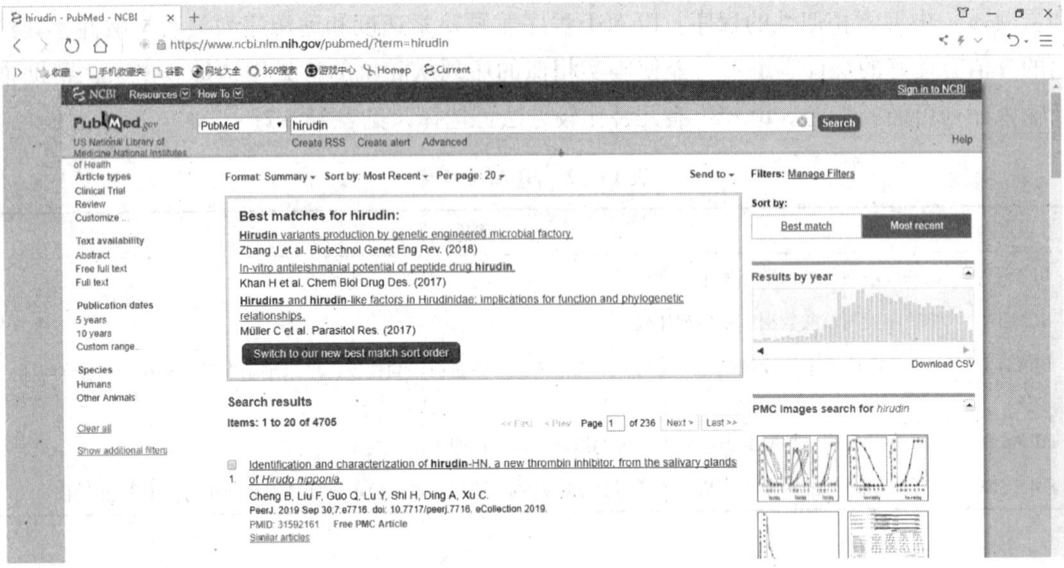

图 18－2　在 PubMed 上输入 *hirudin* 的检索结果

（二）查寻核酸或蛋白质序列

要得到水蛭素基因序列，将 Search 栏下拉菜单选到 Nucleotide，输入关键词 *hirudin*，点击 *Search* 按钮或键盘上 *Enter* 键，检索结果如图 18－3 所示。在列出的条目上寻找与水蛭素基因相关的条目，点击条目上的超链接，得到图 18－4 所示结果。在 Origin 列下可得到核酸序列，其他相关信息可从描述中获得。检索蛋白质序列与检索核酸序列类似，只是在下拉子菜单中需选择 Protein 栏。

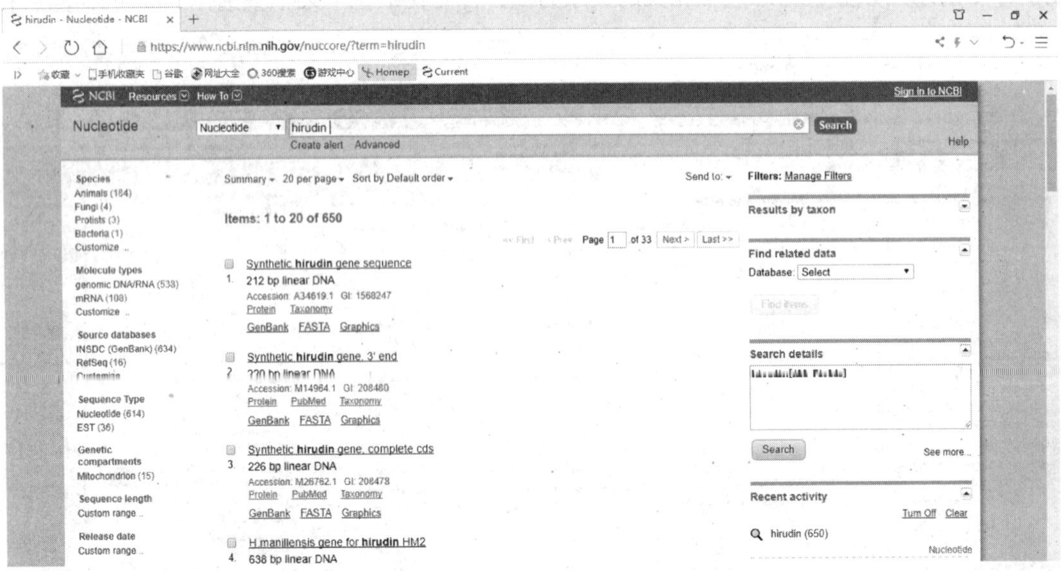

图 18－3　在 Nucleotide 数据库中检索 *hirudin* 核酸序列的结果示例

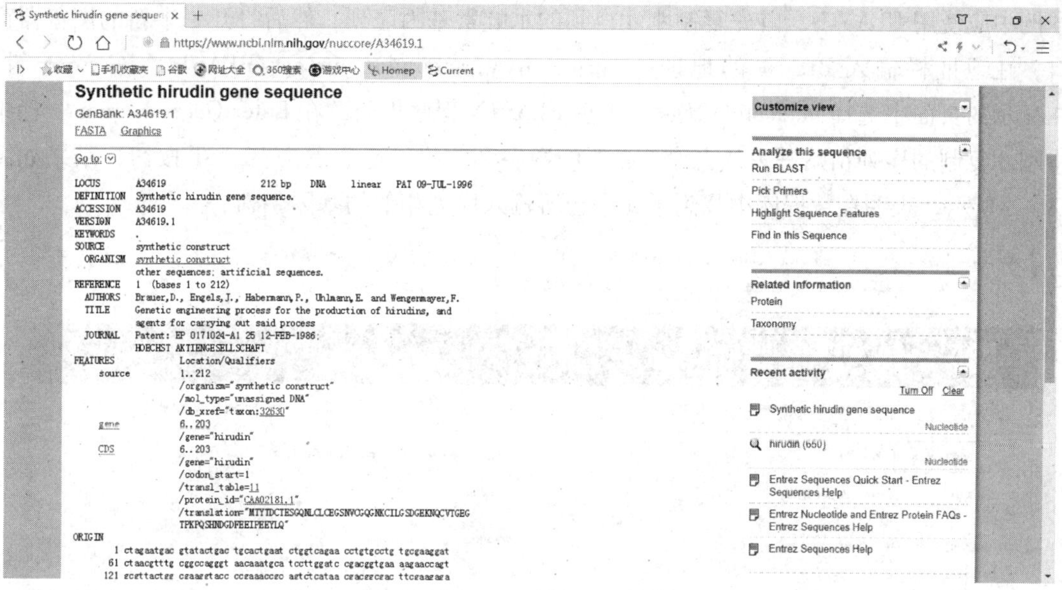

图 18－4　Nucleotide 数据库中 *hirudin* 核酸序列记录示例

（三）检索蛋白质结构

要得到水蛭素的结构，将 Search 栏下拉菜单选到 Structure，输入关键词 *hirudin*，点击 *search* 按钮或 *Enter* 键，检索结果如图 18－5 所示。蛋白质结构测定的主要方法是 X 线衍射和磁共振技术（NMR），在检索结果中列出了所用的结构测定方法，可根据需要选择合适的条目。在结构记录格式上有 PDB 格式和 MMDB 格式，MMDB 结构与原始的 PDB 结构相比，增加了一些附加信息，包括经程序验证的显性化学图像信息、一致的二级结构衍生定义等，但是目前仅有 NCBI 的 Cn3D 软件可以打开和浏览 MMDB 格式的结构，其他软件大部分使用 PDB 格式。下载 PDB 格式的结构文件后，可以使用 Swiss－PdbViewer 打开浏览。

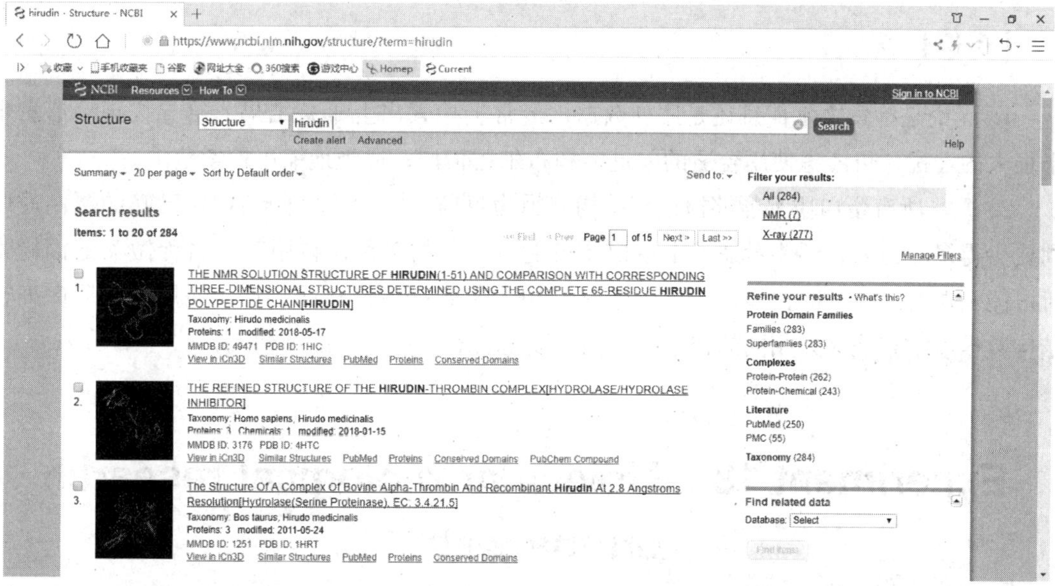

图 18－5　NCBI 的 Structure 数据库中检索 *hirudin* 结果示例

（四）使用 BLAST

如果想要了解在核酸数据库中是否存在与水蛭素相似的基因序列，可使用 BLAST，在本例中应使用 BLASTN。首先复制刚才得到的水蛭素基因序列，然后打开一个新的浏览器窗口，在地址栏输入 BLAST 的地址：http：//www. ncbi. nlm. nih. gov/BLAST/，按 *Enter* 键。在显示的页面中选择 nucleotide blast，出现 BLASTN 程序界面，在 Enter Query Sequence 栏中将刚才复制的序列粘贴进去，去除不必要的数字和空格，然后点击 BLAST 按钮，结果如图 18－6 所示，表明在数据库中找到了 3 条与所输入序列相似的基因序列。

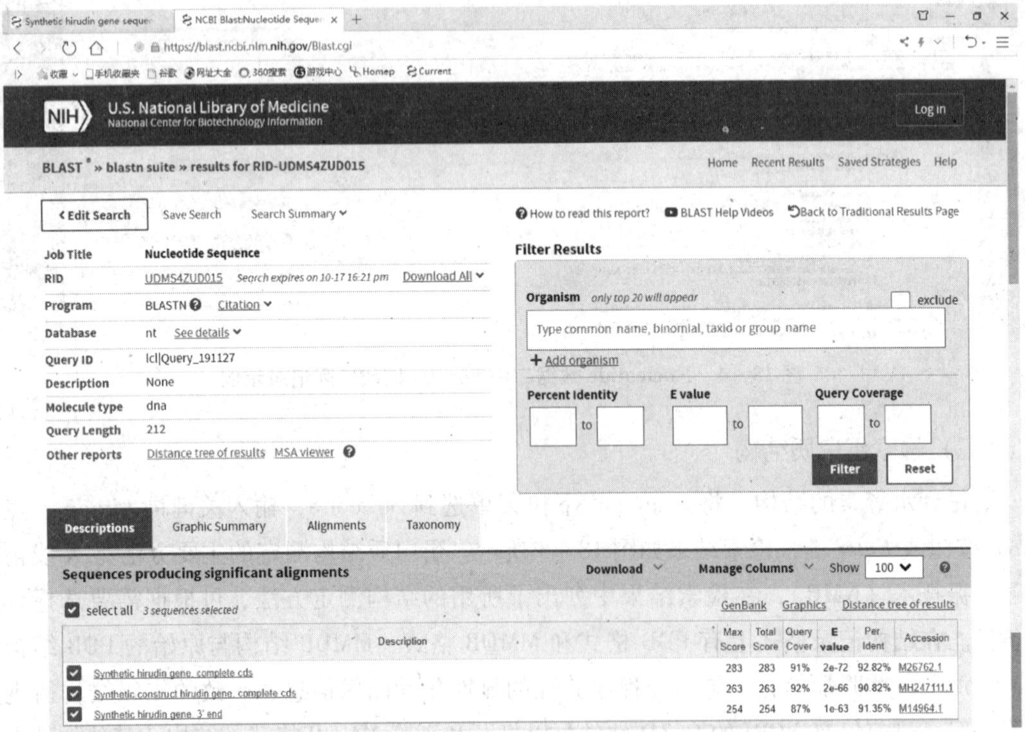

图 18－6　水蛭素核酸序列经 BLASTN 程序检索结果

【讨论】

1. 无论是检索文献信息还是序列数据，经常会检索出上千条结果，给寻找正确信息带来很大的干扰，可考虑使用合适的关键词和布尔逻辑词，适当地缩小检索范围。

2. 并非所有蛋白质都能够检索到结构，因为目前经实验测定的结构比已知的蛋白质序列要少得多。另外，还要考虑一个蛋白质可能有不同的名称，在用某一名称做关键词检索时可能会遗漏。为了防止漏检，可以考虑用 BLAST 数据库搜索的方法，使用蛋白质序列在结构数据库中查找序列相同的记录，从而判断是否有实验测定的结构。

Experiment 18　Molecular biological research on the web

Introduction

The Internet has greatly changed the way researchers search and exchange information. Experimental data can only be printed on paper for communication before, but now it can be obtained from the centralized database in the form of electronic; journals are also published on the Internet in the form of electronic version; with the deepening of genome and post genome research, the collection, storage and analysis of massive biological information has been inseparable from computer and network technology.

The Internet has completely changed the way scientists search for and exchange information. Data that once had to be communicated on paper is now digitized and distributed from centralized databases. Journals are also published online. With the deepening of genome and post genome research, the collection, storage and analysis of massive biological information has been inseparable from computer and network technology.

The Internet has tens of millions of sites at this point. How can we find the useful data? How can we judge which information is useful? In this experiment, we will learn how to use the Internet, and get what we need for bioscience research.

Purpose

1. Master the method of searching nucleic acid and protein sequence.

2. Master the method of retrieving protein structure information and understand the visualization of protein structure.

Principle

1. Database and sequence data retrieval

EMBL, GenBank and DDBJ are the three major DNA sequence databases in the world. EMBL was founded in 1982 by the European Molecular Biology Laboratory. GenBank was established by the National Institute of Health (NIH) in the early 1980s, and then transferred to the National Center for Biotechnology Information (NCBI). DDBJ, the abbreviation of DNA database of Japan, was founded in 1986 and managed by the National Institute of genetics of Japan. In 1988, EMBL, GenBank and DDBJ established the international nucleic acid sequence joint database center and established a cooperative relationship. The three data centers collect sequence data released by sequencing institutions and laboratories, and exchange newly discovered or updated data through computer network every day to ensure the integrity of sequence information of the three databases.

Here we take the database retrieval in NCBI as an example to introduce the retrieval methods of nucleic acid and protein sequence (table 18 – 1).

Table 18 – 1 The main database of NCBI

Database	Introduce
Nucleotide	The Nucleotide database contains sequence data from GenBank, EMBL, and DDBJ, the members of the tripartite, international collaboration of sequence databases.
Protein	The Protein database contains sequence data from the translated coding regions from DNA sequences in GenBank, EMBL and DDBJ as well as protein sequences submitted to Protein Information Resource (PIR), SWISSPROT, Protein Research Foundation (PRF), and Protein Data Bank (PDB) (sequences from solved structures)
Genome	The Genome database provides views for a variety of genomes, complete chromosomes, contiged sequence maps and integrated genetic and physical maps

continued Table

Database	Introduce
Structure and MMDB	The Structure database or Molecular Modeling DataBase(MMDB) contains experimental data from crystallographic and NMR structure determinations. The data for MMDB are obtained from the Protein Data Bank (PDB)
PopSet	The PopSet database contains aligned sequences submitted as a set resulting from a population, a phylogenetic, or mutation study. These alignments describe such events as evolution and population variation. The PopSet database contains both nucleotide and protein sequence data
OMIM	The OMIM(Online Mendelian Inheritance in Man) database is a catalog of human genes and genetic disorders
Taxonomy	The Taxonomy database contains the names of all organisms that are represented in the NCBI genetic database by at least one nucleotide or protein sequence
ProbeSet	ProbeSet database is an Entrez view of NCBI's GEO(Gene Expression Omnibus). Geo is a gene expression and hybridization array repository
3D Domains	3D Domains contains protein domains from the NCBI Conserved Domain Database
UniSTS	A unified, non – redundant view of sequence tagged sites(STSs)
EntrezSNP	A central repository database for both single base nucleotide substitutions and short deletion and insertion polymorphisms
UniGene	An experimental system for automatically partitioning GenBank sequences into a non – redundant set of gene – oriented clusters. Each UniGene cluster contains sequences that represent a unique gene, as well as related information such as the tissue types in which the gene has been expressed and map location.
PubMed	PubMed Central(PMC) is a digital archive that provides access to the full – text of articles from a set of life science journals

Before searching, we need select the database to be searched from the drop – down menu, and then enter the keyword to search. Most search engines use Boolean logic query forms. The Boolean retrieval method limits the retrieval results by connecting the retrieval words with the operators "and", "or", and "not". The meaning of these operators is just like the literal meaning: use "and" to search for documents containing both keywords; use "or" to search for documents containing any (or all) of the two keywords; and "not" to search for documents containing keyword 1 but excluding keyword 2. In addition, the Boolean retrieval is read from left to right just like the text, which can be used as brackets to form more complex retrieval conditions. For example, if you want to retrieve a document with keyword 1, including one of keyword 2 or keyword 3, but without keyword 4, you can enter the following search criteria: [keyword 1 and(keyword 2 or keyword 3)] not keyword 4. You can also use quotation marks to limit a phrase. For example, if you want to retrieve the documents in which the protein sequences appear together in order, you can use "protein sequence" to limit the retrieval results

2. Protein structure retrieval and structure browser

As early as the 1970s, before the emergence of sequence database, Protein Data Bank(PDB) has been established. The PDB database was originally maintained and managed by Brookhaven National Laboratory. In 1998, funded by the National Science Foundation of the United States, the Department of energy and the Institute of health, the research collaboration for structural biology (RCSB) was established, and the PDB database was changed to RCSB. PDB database stores data in the form of text file, and each molecule uses an independent file. In addition to the atomic coordinates, it alsoincludes the source of species, the name of compound, the author and the relevant literature. In addition, resolution, structure factor, temperature coefficient, number of protein main chains, ligand formula, metal ions, secondary structure information, disulfide bond position and other data related to structure are also given. PDB data is stored in text file format, which can be viewed by text editing software, but it is not convenient to view annotation information by text editing software, and it is not able to understand the spatial structure of molecules intuitively.

In the past, imaging of biomacromolecules was impossible. In 1972, Levinthal and Katz created a computer wireframe model. Since then, computer programs have been developed to make the visualization of macromolecular structure, by which the structure can be dragged and rotated, which provides help for computer - aided drug design. Commonly used structural browsers include VRML, Rasmol, chime, etc. Here we only introduce Swiss pdbviewer, which is a free software for protein display and analysis. It provides a large number of menus to meet the needs of viewing protein structure. It can compare the active parts and related structures. The software can be downloaded from the Swiss Institute of bioinformatics (SIB) homepage (https://spdbv. vital - it. ch/).

3. Using BLAST

Basic Local Alignment Search Tool (BLAST) is a set of similarity search programs designed to explore all of the available sequence in databases, regardless of whether the query is protein or DNA. The BLAST programs have been designed for speed, with a minimal sacrifice of sensitivity to distant sequence relationships. The scores assigned in a BLAST search have a well - defined statistical interpretation, making real matches easier to distinguish from random background hits, and help us to found the possible function of query sequence. You can choice different Blast program depending on your query (table 18 - 2).

Tab 18 - 2　BLAST program

Program	Purpose
BLASTP	Find protein sequences similar to query
BLASTN	Find nucleotide sequences similar to query sequence
BLASTX	Find similar proteins to translated query in a protein database, can be used to find the query nucleotide sequence's potential protein products
TBLASTN	Find similar proteins in a translated nucleotide database
TBLASTX	Find similar proteins to translated query in a translated database

Apparatus

PC minimums: CPU Core i3/memory 4GB; Operate system: win32 or higher, with Internet Explorer and link to Internet.

Procedures

1. Search Pubmed

In IE browser, write https://www. ncbi. nlm. nih. gov/ in the address bar, tap Enter key, and the NCBI homepage is shown in Fig. 18 - 1. Here we search for the literature related to hirudin.

First, select Pubmed from the pull – down menu, and then enter the keyword *hirudin* in the dialog box, and click Search or Enter key. The results are shown in Fig 18 –2, indicating that 4705 records related to hirudin are obtained in this search. 20 items are displayed on this page. Boolean expression can be used to narrow the search scope, such as searching for the cloning and expression of hirudin related genes, we can use *hirudin* AND *clone* AND *expression* as the search term. We can try different keywords in the retrieval, so that the retrieval results are more close to our requirements.

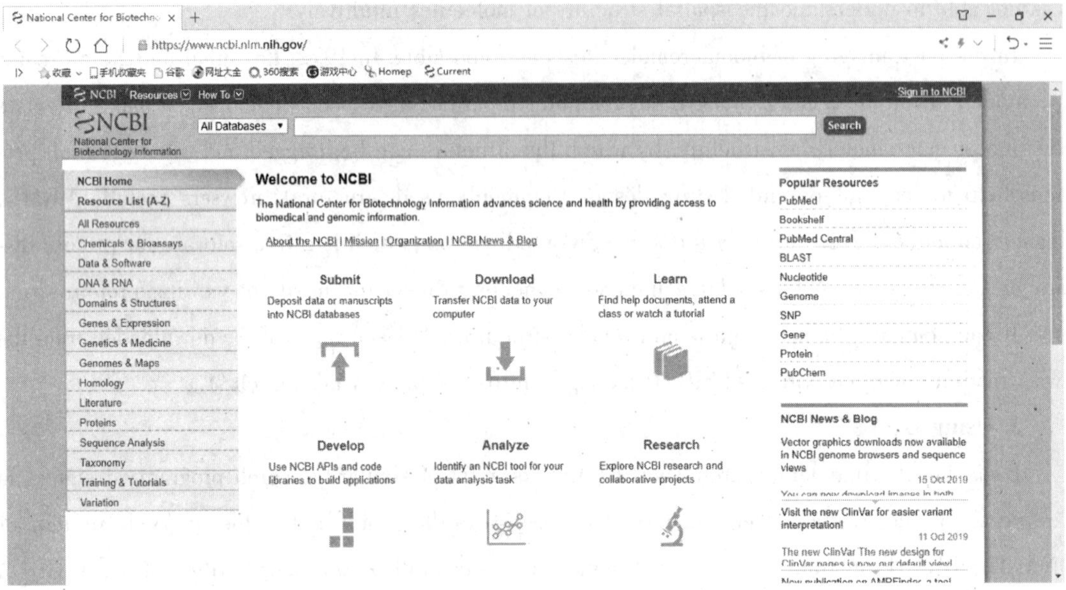

Fig. 18 –1　The homepage of NCBI

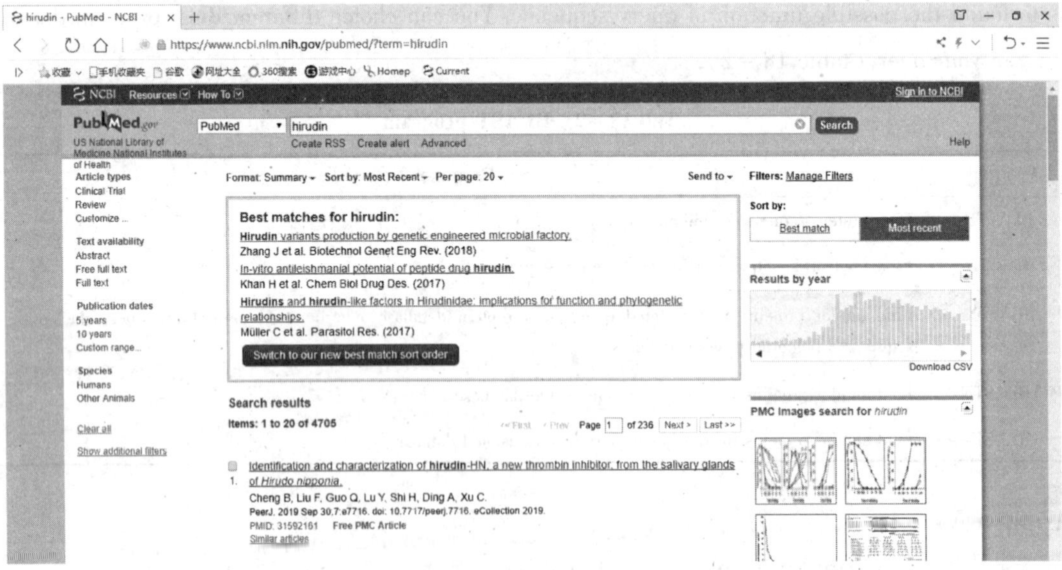

Fig. 18 –2　The result of search *hirudin* in Pubmed

2. Search for nucleotide or protein sequence

To get the nucleotide sequence of hirudin, we select Nucleotide from the pull – down menu, then enter the keyword *hirudin* in the search box, and click Search or Enter key. The result was shown in Fig. 18 –3. Click the hyperlink of hirudin related gene in the list item, and the results are shown in

Fig. 18 – 4. The nucleic acid sequence can be obtained under Origin line, and other relevant information can be obtained from the description. Searching for protein sequence is similar to searching for nucleotide acid sequence, except that you need to select the Protein in the pull – down menu.

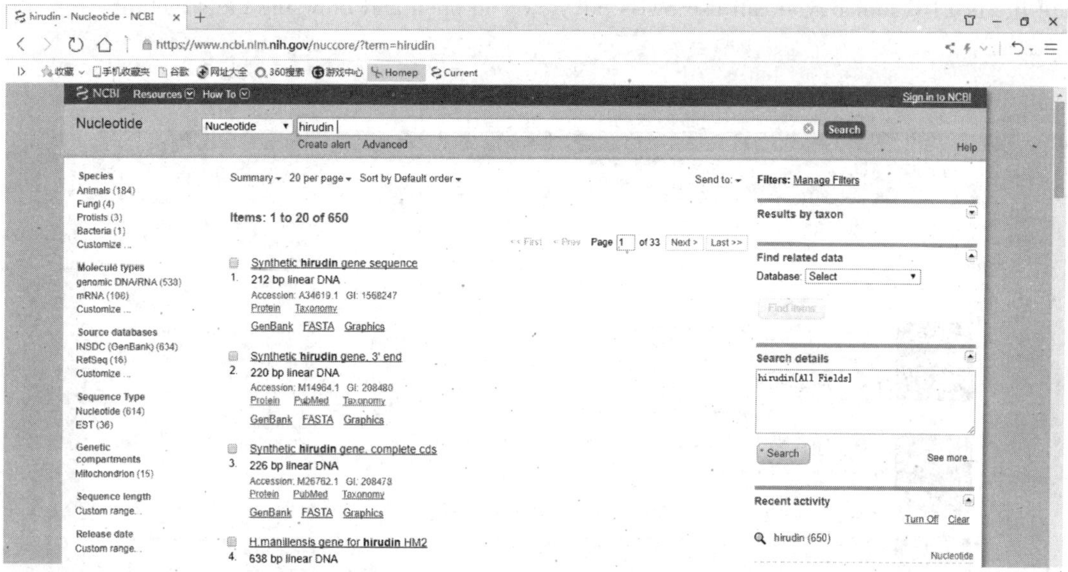

Fig. 18 – 3　example of the result of retrieving *hirudin* nucleotide sequence in nucleotide database

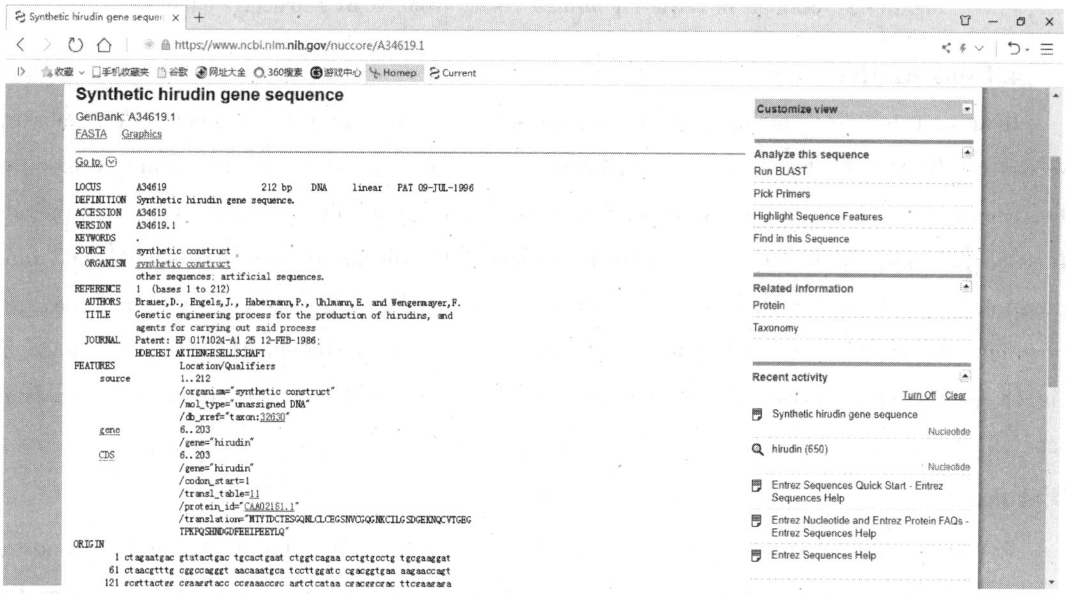

Fig. 18 – 4　example of *hirudin* sequence record in the nucleotide database

3. Search for protein structure

To get the structure of hirudin, select Structure from the pull – down menu in the search column, enter the keyword *hirudin*, and click search or enter. The search results are shown in Fig. 18 – 5. The main methods of protein structure determination are X – ray diffraction and NMR. Therefore, the structure determination methods used are listed in the retrieval results, and appropriate items can be selected according to the needs. There are PDB format and MMDB format in the structure record format. Compared with the original PDB structure, MMDB structure adds some additional information, including explicit chemical image information verified by program, consistent secondary structure derivation definition, etc. , but at present, only Cn3D software of NCBI can open and browse the structure of MMDB format, and most of other software uses PDB format. After downloading the structure file in PDB format, you can use Swiss pdbviewer to open and browsing the structure.

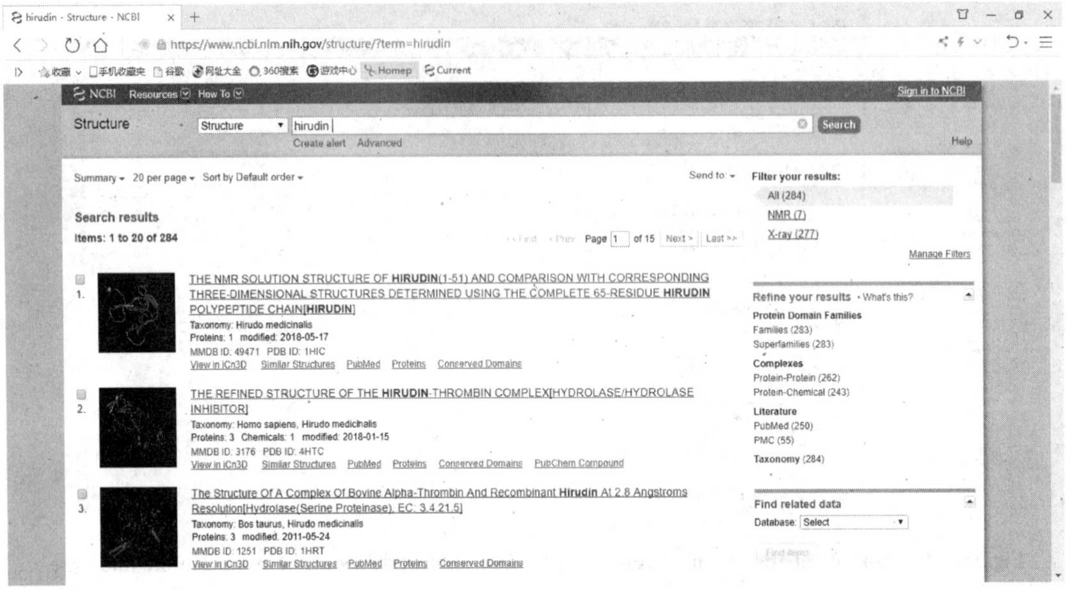

Fig. 18 – 5 example of retrieving *hirudin* results from NCBI's structure database

4. Using BLAST

If we want to know whether there are similar sequences of hirudin in the nucleotide database, we can use BLAST. In this case, we use BLASTN. First, we copy the nucleotide sequence of hirudin. Then we open a new browser window, enter http://www. ncbi. nlm. nih. gov/BLAST/ in the address bar, and press Enter key. Select nucleotide blast in the displayed page, and the BLASTN program interface will appear. Paste the sequence of hirudin in Enter Query Sequence dialog box, remove unnecessary numbers and spaces, and then click the Blast button. The result is shown in Fig. 18 – 6, indicating that three gene sequences similar to the input sequence are found in the database

Discuss

1. Whether it is to retrieve literature information or sequence data, it will often get thousands of results, which make it difficult to the find the correct information. It is recommended to consider different keywords and Boolean logic words, appropriately narrow the results of retrieval, and facilitate the search for correct results.

2. Not all proteins have experimental determined structure, and the search of the structure of a certain protein may get no results. In addition, it should be considered that a protein may have different names, its structure may not be obtain if a different name were used as keywords. In order to avoid the missing, we can use blast program search the same sequence in the structure database, so as to find the experimental structure by the sequence, instead of by the keyword.

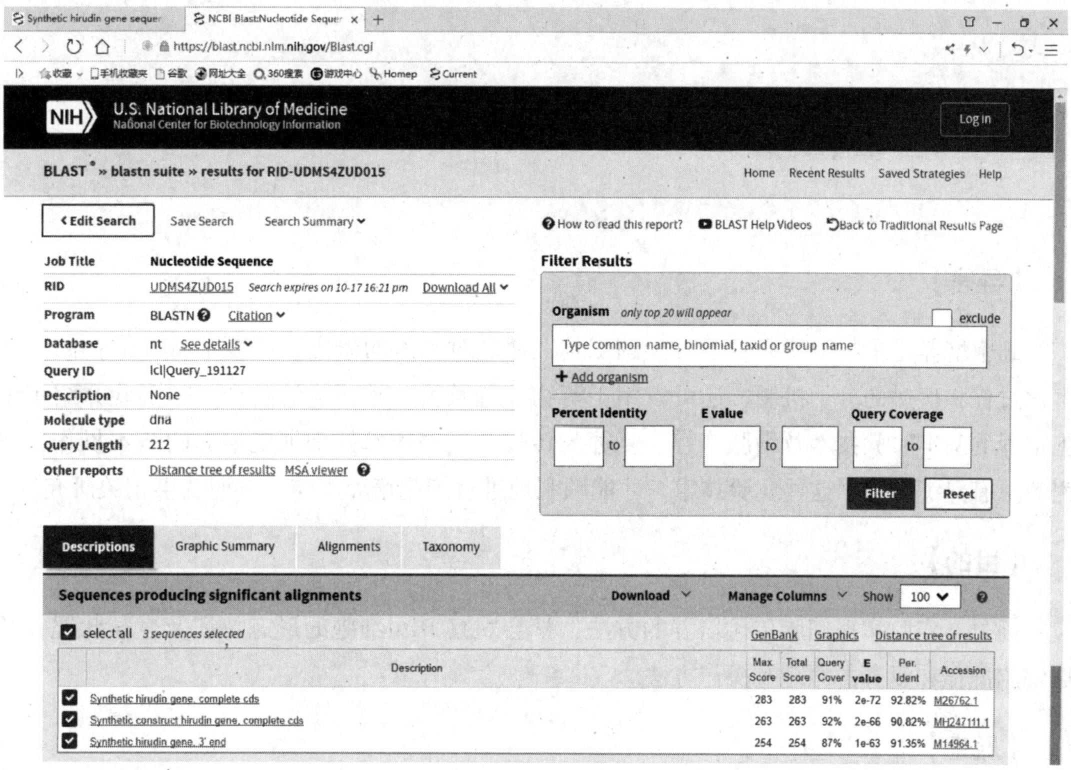

Fig. 18 – 6　search results of hirudin nucleotide sequence by BLASTN program

扫码“学一学”

实验十九　细胞培养基本技术

【导言】

细胞培养是生物学研究必不可少的工具。通过对细胞的研究，可以了解生命体的生理生化过程和疾病的分子机制，还可以用于药物筛选及蛋白类药物的生产。本实验介绍了细胞培养的基本实验技术及细胞冻存、复苏与传代。这些技术都是细胞培养的基本操作，研究者只有熟练掌握了这些实验技术，才能顺利地进行细胞培养工作，进而从事相关研究。

【目的】

通过本实验掌握细胞传代培养的方法，学会观察 B16 细胞的形态特征及生长状况，掌握细胞冻存和复苏的原理及操作方法。

【原理】

（一）细胞的传代培养

体外培养的细胞生长一段时间后，由于密度增大会出现生存空间不足以及营养枯竭。为了将细胞密度维持在最佳水平，以便在体外持续地培养，需要对细胞进行传代培养，即将培养的细胞分散，从容器中取出，以一定的比例进行稀释，转移到另外的容器中进行培养。

体外细胞培养按照细胞生长状态可分为悬浮培养和贴壁培养。悬浮培养是细胞在培养基中悬浮生长。悬浮培养细胞传代比较简单，收集细胞，离心去除旧培养基，加入新鲜培养基，按一定比例稀释分散，转移至另外容器中培养即可。贴壁培养的目的是使细胞附着在介质上单层生长。许多细胞都具有贴壁依赖性，而贴壁生长的细胞传代过程也比较复杂。本实验以贴壁生长细胞为例，介绍细胞的传代培养过程。

（二）细胞的冻存和复苏

连续细胞培养可能导致细胞基因变异，还可能增加细胞污染的可能性，因此细胞的冻存对保种来说非常重要。对于新获得的细胞系要及时冻存保种。目前细胞冻存多将细胞分散于完全培养基中，采用甘油或二甲基亚砜作保护剂。冷冻保护剂能提高细胞膜对水的通透性，降低冰点，加上缓慢冷冻可减少冻存过程中细胞内冰晶的形成，从而减少细胞损伤。冻存细胞需要复苏后才能重新使用。细胞复苏的过程对细胞有伤害。缓慢融化会造成水分渗入细胞和再结晶，损伤细胞。复苏细胞应采用快速融化的方法，以确保细胞的高存活率。

【试剂与器材】

（一）细胞的传代培养

1. 肿瘤细胞 B16。

2. 细胞培养基 RPMI 1640。
3. 胎牛血清。
4. 0.25% 胰蛋白酶溶液。
5. 磷酸盐缓冲液（PBS）。
6. 细胞培养瓶。
7. 离心管。

（二）细胞的冻存和复苏

1. 肿瘤细胞 B16。
2. 细胞培养基 RPMI 1640。
3. 胎牛血清。
4. 胰蛋白酶。
5. 磷酸盐缓冲液 PBS。
6. DMSO。
7. 细胞培养瓶。
8. 离心管。
9. 冻存管。

【实验步骤】

（一）细胞的传代培养

扫码“看一看”

1. 从培养箱中取出细胞，在显微镜下观察细胞形态。
2. 在生物安全柜内，打开细胞培养瓶，弃去细胞旧培养液。
3. 加入 PBS 溶液轻轻润洗细胞。为避免影响细胞，洗涤液应加在培养瓶没有细胞的一侧。轻轻晃动培养瓶润洗细胞，弃去 PBS。
4. 加入预热的 0.25% 胰酶，其体积应覆盖细胞层。
5. 室温或 37℃孵育，不时在显微镜下观察细胞。可以轻轻敲打培养瓶帮助细胞解离。
6. 当细胞刚开始脱落时（形态变成圆形），加入 3～4ml 预热的完全培养液（含 10% FCS），吹打成单细胞悬液。
7. 将细胞转移至离心管内，800～1000rpm 离心 10 分钟，弃去上清液。
8. 重新悬浮细胞，取样进行细胞计数（可用细胞计数板或细胞计数仪）。
9. 计算细胞数目，在细胞培养瓶中加入培养基，将细胞稀释至适合的浓度。
10. 在细胞瓶上标注细胞名称、代数、日期以及操作人姓名等信息，放入培养箱继续培养。

（二）细胞冻存

1. 配制含 10% DMSO 或甘油、10%～20% 血清的冻存培养液。冻存培养液冷藏保存待用。
2. 取对数生长期的细胞，按照细胞传代的方法消化收集重悬细胞。

扫码“看一看”

3. 用细胞计数板进行细胞计数，台盼蓝染色计算细胞存活率。
4. 800～1000rpm 离心 10 分钟，去除上清。
5. 加入配制好的冻存培养液，用吸管轻轻吹打使细胞均匀，调整冻存液中细胞的最终

密度为 5×10^6/ml ~ 1×10^7/ml。将细胞分装入冻存管中，每管 1 ~ 1.5ml。

6. 在冻存管上标明细胞的名称、代数、冻存时间及操作者姓名。

7. 标准冻存程序降温速率为 1 ~ 2℃/min；当温度在 -25℃以下时，可增至 5 ~ 10℃/min；到 -100℃时，则可迅速浸入液氮中。实际操作中可将细胞放入专门的细胞冻存盒中，然后放入 -70℃冰箱中过夜，第二天取出冻存管，移入液氮容器内长期保存。也可将装有细胞的冻存管在 4℃放置 40 分钟后放入 -20℃冰箱 1 小时，然后放入 -70℃冰箱中过夜，第二天取出冻存管，移入液氮容器内保存。

（三）细胞复苏

扫码"看一看"

1. 从液氮容器中取出冻存管，立即浸入 37℃温水中。

2. 轻轻摇动令其尽快融化，当管中还有一点冰块时，从水浴中取出冻存管，移至超净台内，乙醇擦拭冻存管外壁。

3. 小心将解冻细胞转移至具有 5ml 细胞完全培养基的离心管内，轻柔混匀。

4. 800 ~ 1000rpm 离心 5 分钟。

5. 弃去上清液，加入含 10% 血清的培养液重悬细胞，接种培养瓶，37℃培养箱静置培养。

6. 次日更换一次培养液，继续培养。

【注释】

1. 细胞培养的必要条件具体如下。

（1）培养基　是细胞培养中最重要的成分，可以为细胞提供必要的营养成分。基础培养基中含有氨基酸、维生素、无机盐和碳源。另外，含有的缓冲体系可以调节培养体系的 pH 和渗透压。一般基础培养需添加一定浓度的血清后使用。

（2）血清　其中含有细胞生长所需的各种生长因子和黏附因子、激素、脂类等物质。

（3）适宜的 PH　大多数哺乳动物细胞的最适 pH 为 7.4，而有些昆虫细胞的最适 pH 为 6.2。

（4）适宜的温度　大多数哺乳动物细胞的最适温度为 37℃，而昆虫细胞在 27℃呈现最佳生长状态。

（5）二氧化碳　培养基中通常添加缓冲盐体系，如二氧化碳 - 碳酸氢盐体系，来缓冲 pH 的变化。由于培养基的 pH 取决于溶解态二氧化碳和碳酸氢盐的精密平衡，空气中二氧化碳含量的变化会影响培养基的 pH。所以需要使用二氧化碳培养箱来提供外源二氧化碳。一般二氧化碳的浓度为 5%。

2. 细胞培养所用材料均需保持无菌，实验过程要遵循无菌操作，防止细胞被细菌、真菌、支原体或病毒污染。

3. 贴壁细胞汇合度达 80% ~90% 即可进行传代。贴壁细胞完全汇合时会产生接触抑制而停止生长，有些细胞会发生性质变化。

4. 使用胰蛋白酶消化细胞应严格控制消化时间。消化时间太短，细胞难以脱落；消化时间过长，会导致细胞受损。可使用含 10% 血清的培养基终止消化过程。

5. 冻存时细胞浓度要高，而且要尽可能在细胞传代数较低时冻存细胞。

6. 为了保证冻存效果，可在冻存前检测细胞活力，细胞存活率应在 90% 以上。可用台盼蓝染色计算细胞存活率。将细胞和台盼蓝染料混合 3 分钟，用细胞计数板在显微镜下观察。活细胞为无色，死细胞被染成蓝色。计数并计算细胞存活率。

7. 细胞冻存时要逐渐降低温度，也可使用可控制降温速度的低温冷冻容器。冻存时必须使用含有 DMSO 或甘油等冻存保护剂的冻存培养基。

8. 细胞复苏时要迅速，解冻尽量在 1 分钟内完成。

9. 细胞复苏时要高密度接种，以提高细胞的复苏效果。

Experiment 19　Basic cell culture techniques

Introduction

Cell culture is one of the indispensable tools for biology research. Cell culture can be used to study the physiology and biochemistry of cells, the mechanism of diseases. It is also used in drug screening and large scale manufacturing of pharmaceutical proteins. This chapter introduces the basic experimental techniques of cell culture, including cell subculture, cell cryopreservation and cell thawing. These are basic techniques in cell culture. Only by mastering these experimental techniques skillfully, can researchers carry out cell culture work smoothly, and then engage in related research.

Purpose

Through this experiment, we master the method of cell subculture, learn to observe the morphological characteristics and growth status of B16 cells, and master the principle and operation method of cell cryopreservation and resuscitation.

Principle

1. Subculture Adherent Cells

The cells cultured in vessels proliferate exponentially and this will lead to inadequate living space and nutrient exhaustion. To keep them at an optimal density for continued growth, it is necessary to subculture cells. That means the cultured cells will be dispersed, removed from the container, diluted in a certain proportion, and transferred to another containers for culture.

There are two main cell culture systems, adherent culture (asmonolayers on an artificial substrate) or suspension culture (floating in the culture medium). For suspension cells, the passage procedure is relatively easier requiring collecting and diluting cells. For adherence culture, cells have to be dissociated from the substrate. Most cells are adherent – dependent and the process of passage for adherent culture is more complex. Therefore, in this experiment, adherent culture is taken as an example to introduce the method of cell subculture.

2. Cryopreservation and thawing

Continuous culture of cell leads to accumulation of unwanted genetic alterations. In addition, continuous cultureincreases the possibility of contamination. It is important that the cells are cryopreserved for seed stock. For newly acquired cell lines, cryopreservation of stock vials should be done as soon as possible.

The best way forcryopreserving cells is to store them in liquid nitrogen in complete medium with a cryoprotective agent such as glycerol or dimethylsulfoxide(DMSO). Cryoprotective agents increase permeability of cell membrane, decrease the freezing point. With a slower cooling rate, the risk of ice crystal formation, which can damage cells and cause cell death, can be reduced.

Preparing a fresh cell culture from cryopreserved stocks needs thawing. The thawing procedure is harmful to frozen cells. Slow melting causes water infiltration and recrystallization, which leads to cell damage. Using correct method and working quickly ensures a high cells survival proportion.

Reagents and apparatus

1. Subculture Adherent Cells

(1) B16 cells.

(2) cell culture medium RPMI 1640.

(3) fetal bovine serum.

(4) 0.25% Trypsin.

(5) Phosphate Buffered Saline(PBS).

(6) cell culture flask.

(7) 15ml centrifuge tubes.

2. Cryopreservation and thawing

(1) B16 tumor cells.

(2) Cell culture medium RPMI 1640.

(3) Fetal bovine serum.

(4) Trypsin.

(5) Phosphate buffered saline(PBS).

(6) DMSO.

(7) Cell culture flask.

(8) Centrifuge tube.

(9) Cryogenic vials.

Procedure

1. Subculture ddherent cells

Prior to beginning any cell culture work, the biological safety cabinet should be turned on and allowed to run for 30min to purge the contaminated air. All work surfaces within the cabinet should be decontaminated with 70% ethanol. Any materials required for the procedure should be decontaminated and placed in or near the cabinet.

(1) Take out the cell flask and view the cells under microscope.

(2) In bio-safety cabinets, open the cell flask and remove the cell culture media from the culture vessel.

(3) Wash cells with PBS. Add wash solution gently to the side of the flask opposite the attached cell layer to avoid disturbing the cell layer, and shake the vessel several times. Remove and discard the wash solution from the culture vessel.

(4) Add the pre-warmed 0.25% trypsin into the flask. The solution should cover the cell layer.

(5) Incubate the culture vessel in the 37℃ incubator or at room temperature for approximately 2-3minutes, and check the cells under the microscope for detachment. You may also tap the vessel to help cell detachment.

(6) When the cells become just detached (cells become round), add 3-4ml new pre-warmed culture medium (with 10% FCS) into the vessel. Disperse the cells by pipetting over the cell layer surface several times.

(7) Transfer the cells to a sterilized tube and centrifuge cells at 800-1000rpm for 10minutes. Discard the supernatant.

(8) Resuspend the cell pellet in a minimal volume of culture medium and take a sample for counting. Using a hemacytometor or a cell counter.

(9) Determine the total number of cells. Dilute cell suspension to a desired cell concentration in a culture vessel.

(10) Mark the cell name, passage number, date and operator name on the cell flask, then put the cells inincubator for continued culture.

2. Cryopreservation cells

(1) Prepare freezing medium: RPMI1640 with 10%-20% FCS and 10% DMSO. Keep the freezing medium at 4℃ until use.

(2) Gently detach cells from the tissue culture flask and collect the cells following the guidelines of cell subculture.

(3) Determine the cell number and viability using a hemacytometer and Trypan Blue exclusion.

(4) Centrifuge the cells at approximately 800-1000rpm for 10minutes and remove the supernatant without disturbing the cell pellet.

(5) Resuspend the cell pellet in cold freezing medium at the desired cell density (5×10^6/ml-1×10^7/ml) and dispense aliquots of the cells into cryogenic storage vials.

(6) Mark the cell name, passage number, date, and operator name on the vials.

(7) The standard freezing procedure is decreasing the temperature approximately 1℃ per minute. When the temperatureis below -25℃, temperature can be decreased by 5-10℃ per minute. When the temperature reaches -100℃, the cells can be put in liquid nitrogen for long-term storage. In practice, cells can be put in a controlled rate freezing apparatus and keep at -80℃ overnight. Thc next day, transfer frozen cells to liquid nitrogen for long-term storage. Alternatively, cells can be placed at 4℃ for 30min, then at -20℃ for 1hour and then at -80℃ overnight. The next day, transfer frozen cells to liquid nitrogen.

3. Thawing cells

(1) Take thecryovial containing the frozen cells from liquid nitrogen and place it into a 37℃ water bath immediately.

(2) Quickly thaw the cells by gently shaking the vial in the 37℃ water bath until there is just a small bit of ice left in the vial. Transfer the vial into biological safety cabinet. Before opening, steri-

lize the outside if the vial with 70% ethanol.

(3) Transfer the thawed cells to a centrifuge tube containing 5ml of pre - warmed cell culture medium and mix the cells by gently shaking.

(4) Centrifuge the cell suspension at approximately 800 - 1000rpm for 5minutes.

(5) Remove the supernatant carefully without disturbing the cell pellet. Gently resuspend the cells in complete growth medium, and transfer them into the appropriate culture vessel. Culture the cells in a CO_2 incubator.

(6) Change the culture medium the next day and culture the cells continually.

Notes

1. Cell culture conditions vary widely for each cell type, but the following factors are invariably necessary for all cell cultures.

(1) Culture medium: Basal media contain amino acids, vitamins, inorganic salts, and a carbon source such as glucose. It also include buffering system to control the pH and osmotic pressure. Basal media usually have to be supplemented with serum.

(2) Serum: Serum is derived from blood and it provides important source of growth and adhesion factors, hormones, lipids and minerals.

(3) pH Levels: The optimum pH is 7.4 for most mammalian cells, but 6.2 for some insect cells.

(4) Temperature: Most mammalian cells are maintained well at 37℃, but the optimum temperature for some insect cells is 27℃.

(5) CO_2 Levels: The growth medium usually contain a buffering system, e. g. carbon dioxide (5) (CO_2) - bicarbonate (HCO^{3-}) based buffer. Because the pH of the medium is dependent on the delicate balance of dissolved carbon dioxide and bicarbonate, it is necessary to use exogenous CO_2. 5% CO_2 in air is recommended.

2. To avoid any contamination of bacteria, molds, mycoplasma or viruses, all the materials required for cell culture should be decontaminated and the procedure should follow aseptic technique.

3. The adherent cells should besubcultured when the confluency reaches 80% - 90%. If cells are allowed to reach 100% confluency, contact inhibition can result growth arrest with a decrease in cell proliferation. Some cells may have genetic changes.

4. When trypsin is used to dissociate the cells, the operator should monitor the time carefully to avoid cell damage. Cell medium with 10% serum can be used to stop the dissociation.

5. Freeze cells at a high concentration and at as low a passage number as possible.

6. To get a better freezing result, check the the cell viability before freezing. The cell viability should be above 90%. Trypan blue can be used for cell viability checking. Briefly, mix the cells with trypan blue and check the cells using a hemocytomete under the microscope. Count the numbers of bright clear (viable) and nonviable (blue) cells. Calculate the percentage of viable cells.

7. Freeze the cells slowly. Cryoprotective agent must be used in freezing medium. It is recommended to use a controlled rate freezer.

8. Thaw frozen cells rapidly (< 1minute) in a 37℃ water bath.

9. Culture thawed cells at high density.

实验二十　免疫细胞分离技术

扫码“学一学”

扫码“看一看”

【导言】

用体外方法对机体各种具有免疫反应的细胞分别做鉴定、计数和功能测定，是观察机体免疫状态的一种重要手段。为此，须将各种参与免疫反应的细胞从血液或脏器中分离出来。参与免疫反应的细胞主要包括淋巴细胞、巨噬细胞、中性粒细胞等。由于检测的目的和方法不同，分离细胞的需求和技术也异。有的仅需要白细胞，有的则需要分离单个核细胞（monouclear cell），其中含淋巴细胞和单核细胞（monocyte），有的则需分离T细胞和B细胞以及其亚群。分离细胞选用的方法应力求简便可行，并能获得高纯度、高获得率、高活力的细胞。现用分离细胞群的原则，一是根据各类细胞的大小、沉降率、黏附和吞噬能力加以区分；二是按照各类细胞的表面标志，包括细胞表面抗原和受体，加以选择性分离。

【目的】

通过本实验了解淋巴细胞分离的工作原理和操作步骤。

【原理】

（一）Ficoll密度梯度离心法

外周血单个核细胞（peripheral blood mononuclear cell，PBMC）是血液中具有单个核的细胞，主要包括淋巴细胞、单核细胞、吞噬细胞、树突状细胞和少量其他类型细胞，其中淋巴细胞占很大一部分。分离PBMC的主要目的是去除多核细胞和红细胞，从而能够方便地模拟体外的血液免疫环境。

Ficoll是蔗糖的多聚体，呈中性，平均分子质量为400 000，当密度为1.2g/ml时，仍未超出正常生理性渗透压，也不穿过生物膜。红细胞、粒细胞比重大，离心后沉于管底；淋巴细胞和单核细胞的比重小于或等于分层液比重，离心后漂浮于分层液的液面上，也可有少部分细胞悬浮在分层液中。吸取分层液液面的细胞，就可从外周血中分离到PBMC。

（二）免疫磁珠法

免疫磁珠法分离细胞是基于细胞表面抗原能与接有磁珠的特异性单抗相结合这一原理。在外加磁场中，通过抗体与磁珠相连的细胞被吸附而滞留在磁场中，而无该种表面抗原的细胞由于不能与接有磁珠的特异性单抗结合而没有磁性，不在磁场中停留，从而使细胞得以分离。免疫磁珠法分为阳性分离法和阴性分离法。①阳性分离法：磁珠结合的细胞就是所要分离获得的细胞。②阴性分离法：磁珠结合不需要的细胞，游离于磁场的细胞为所需细胞。一般而言，阴性分离法的磁珠用量比阳性分离法大。

【试剂与器材】

（一）Ficoll密度梯度离心法

1. 比重1.077±0.001的聚蔗糖－泛影葡胺聚蔗糖（商品名为淋巴细胞分离液）。

2. Hank's 液（无 Ca^{2+} 和 Mg^{2+}）。

3. 10% 小牛血清 RPMI 1640。

4. 0.2% 台盼蓝染色液，用生理盐水或等渗的 PBS 配制。

5. 肝素用 Hank's 液或生理盐水稀释成 500 单位/毫升，抗凝人外周血肝素用量约为 50 单位/毫升血。

6. 离心管、毛细滴管、1ml 和 10ml 刻度吸管。

7. 无菌干燥注射器针头。

8. 碘酒，75% 乙醇，无菌棉球，镊，橡皮止血带。

9. 细胞计数板、水平式离心机。

（二）免疫磁珠法分离 $CD4^+T$ 细胞

1. 生物素化的人 $CD4^+$ 阴性分选抗体混合物［cocktail，内含抗 B 细胞（CD45R 和 B220）、$CD8^+T$ 细胞（CD8a 和 Ly－2）、造血细胞（CD11b 和 Mac－1）及 NK 细胞（CD49b 和 DX5）等非 $CD4^+T$ 细胞表面标志的抗体］。

2. 结合磁珠的抗生物素抗体。

3. 含 0.5% BSA（或 0.5% FCS）及 2mmol/L EDTA 的 PBS 缓冲液。

4. 磁珠分离器或分离柱。

【实验步骤】

（一）Ficoll 密度梯度离心法分离淋巴细胞

1. 在离心管中加入适量淋巴细胞分离液。

2. 取肝素抗凝静脉血与等量 Hank's 液或 RPMI 1640 充分混匀，用滴管沿管壁缓慢叠加于分层液面上，注意保持清楚的界面。水平离心 2000rpm×20 分钟。

3. 离心后管内分为 3 层：上层为血浆和 Hank's 液，中层为淋巴细胞分离液，下层主要为红细胞和粒细胞。在上、中层界面处有一以单个核细胞为主的白色云雾层狭窄带，此外，还含有血小板（图 20－1）。

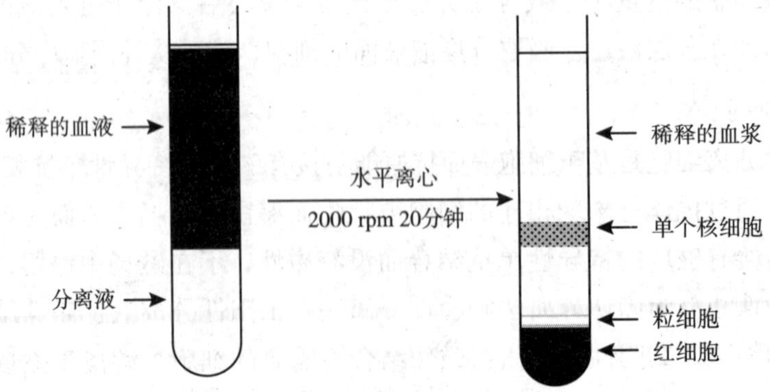

图 20－1　Ficoll 密度梯度离心后分层示意

4. 用毛细血管插到云雾层，吸取单个核细胞。置入另一离心管中，加入 5 倍以上体积的 Hank's 液或 RPMI 1640，1500rpm×10 分钟，洗涤细胞两次。

5. 末次离心后，弃上清，加入含有 10% 小牛血清的 RPMI 1640，重悬细胞。取一滴细胞悬液与一滴 0.2% 台盼蓝染液混合，于细胞计数板上计数 4 个大方格内的细胞总数。

（二）免疫磁珠法分离 $CD4^+$T 细胞

1. 将已分离出的 PBMC 用 PBS 缓冲液洗涤两次，重悬。
2. 加入生物素化的、不针对 $CD4^+$T 细胞的抗人抗体混合物，4℃孵育 10 分钟。
3. PBS 洗涤两次、重悬。
4. 加入 PBS 缓冲液及结合磁珠的抗生素抗体。4℃孵育 15 分钟。
5. PBS 洗涤两次、重悬。
6. 用 MACS 仪器选择适当的分选程序，取阴性部分，即为 $CD4^+$T 细胞。

【注释】

1. 全血溶液可以加在 Ficoll 上层或下层，但是最终必须保证两种溶液分层清晰。
2. 分离 PBMC 第一步离心时，一定不能设置制动，或设置低水平的制动，否则分层混乱。
3. 抗体包被磁珠与死亡细胞和红细胞常有非特异性结合，所以分选前去除死亡细胞和红细胞可以提高分选效率。
4. 分选细胞用作培养时，全过程注意无菌培养。

Experiment 20　Isolation of immune cells

Introduction

The identification, counting and functional determination of all kinds of immune response cells in vitro is an important means to observe the immune state of the body. To this end, a variety of cells involved in the immune response must be isolated from the blood or organs. The cells involved in immune response mainlyinclude lymphocytes, macrophages, neutrophils and so on. Due to the different purposes and methods of detection, the needs and techniques of cell isolation are also different. Some only need white blood cells, some need to isolate mononuclear cell, which contains lymphocytes and monocytes, while others need to isolate T and B cells and their subsets. The method of cell isolation should be simple and feasible, and the cells with high purity, high yield and high vitality can be obtained. At present, the principle of separating cell groups is to be composed according to the size, sedimentation rate, adhesion and phagocytosis of all kinds of cells, and the other is to selectively separate according to the surface markers of all kinds of cells, including antigens and receptors on the surface of cells.

Purpose

Through this experiment, we understand the working principle and operation steps of lymphocyte separation.

Principle

1. Ficoll density gradient centrifugation

PBMC(peripheral blood mononuclear cell), peripheral blood mononuclear cells, as the name implies, its main cell type is the blood mononuclear cells, mainlyincluding lymphocytes, mononuclear cells, phagocytes, dendritic cells and a small number of other cell types. Lymphocytes account for a large part of it. The main purpose of isolating PBMC is to remove multinucleated cells and red blood cells, so that it can easily simulate the blood immune environment *in vitro*.

Ficoll is a sucrose polymer with an average molecular weight of 400 000. When the density is 1.2g/ml, it does not exceed the normal physiological osmotic pressure and does not pass through thebiofilm. The specific gravity of red blood cells and granulocytes is large, and the specific gravity of lymphocytes and monocytes is less than or equal to the specific gravity of the stratified solution. After centrifugation, a small number of cells can be suspended in the stratified solution. By absorbing the cells on the liquid level of stratified fluid, mononuclear cells can be isolated from peripheral blood.

2. Immunomagnetic bead

The separation of cells by immunomagnetic beads is based on the fact that the cell surface antigen can bind to the specific monoclonal antibody connected to the magnetic beads. In the external magnetic field, the cells connected to the magnetic beads by antibodies are adsorbed and retained in the magnetic field. Cells without the surface antigen are not magnetic because they cannot bind to the specific monoclonal antibody attached to the magnetic beads and do not stay in the magnetic field, thus isolating the cells.

Immunomagnetic beads method can be divided into positive separation method and negative separation method: ①The positive separation method: the cells bound by magnetic beads are the cells to be isolated. ②The negative separation: magnetic beads bind unwanted cells, and cells dissociated from the magnetic field are the required cells.

Generally speaking, the amount of magnetic beads in the negative separation method is larger than that in the positive separation method.

Reagents and apparatus

1. Ficoll density gradient centrifugation

(1) Specific gravity 1.077 ± 0.001 polysucrose – diatrizoatepolysucrose (trade name: lymphocyte separation solution).

(2) Hank's solution (no Ca^{2+} or Mg^{2+}).

(3) 10% calf serum RPMI 1640.

(4) 0.2% Trypan blue staining solution, prepared with normal saline or isotonic PBS.

(5) Heparin is diluted into 500units/ml, anticoagulant human peripheral blood heparin with Hank's solution or normal saline. The dosage of heparin is about 50units/1ml of blood.

(6) Centrifugal tubes, capillary droppers, 1ml and 10ml calibration straws.

(7) Aseptic drying syringe needle.

(8) Iodine, 75% alcohol, sterile cotton balls, tweezers, rubber tourniquet.

(9) Blood cell counting board, microscope, horizontal centrifuge.

2. $CD4^+$ T cells were isolated by immunomagnetic beads

(1) Biotinylated human $CD4^+$ negative sorting antibody mixture [cocktail, contains antibodies against B cells (CD45R, B220), $CD8^+$ T cells (CD8a, Ly - 2), hematopoietic cells (CD11b, Mac - 1), NK cells and other non - $CD4^+$ T cell surface markers].

(2) Anti - biotin antibody (antibody) bound to magnetic beads.

(3) PBS buffer containing 0.5% BSA (or 0.5% FCS) and 2mmol/L EDTA.

(4) Magnetic bead separator or separation column.

Procedures

1. Isolation of lymphocytes by Ficoll density gradient centrifugation

(1) An appropriate amount of lymphocyte separation solution was added to the centrifuge tube.

(2) The heparin anticoagulant venous blood was fully mixed with the same amount of Hank's solution or RPMI 1640, and the dropper was slowly superimposed on the stratified liquid surface along the tube wall. Horizontal centrifugation 2000rpm × 20minutes.

(3) After centrifugation, the tube was divided into three layers, the upper layer was plasma and Hank's solution, and the lower layer was mainly red blood cells and granulocytes. The middle layer is the lymphocyte separation solution, in the upper and middle interface there is a narrow band of white cloud layer dominated by mononuclear cells, mononuclear cellsincluding lymphocytes and monocytes (Fig. 20 - 1). In addition, it also contains platelets.

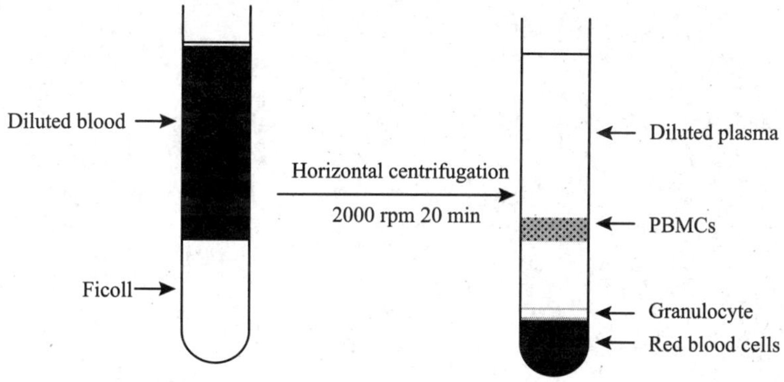

Fig. 20 - 1　Ficoll density gradient centrifugation

(4) Insert capillaries into the cloud layer to absorb mononuclear cells. Put into another centrifugal tube, add more than 5times the volume of Hank's solution or RPMI 1640, 1500rpm × 10minutes, wash the cells twice.

(5) After the last centrifugation, the supernatant was discarded and RPMI 1640, cells containing 10% calf serum were added. A drop of cell suspension was mixed with a drop of 0.2% trypan blue dye, and the total number of cells in the four squares was counted on the blood cell count board.

2. Isolation of $CD4^+$ T cells by immunomagnetic beads

(1) The separated PBMC was washed twice with PBS buffer and resuscitated.

(2) A mixture of biotinylated anti - human antibodies not targeting $CD4^+$ T cells was added and

incubated at 4℃ for 10minutes.

(3) PBS is washed twice and resuscitated.

(4) Add PBS buffer and antibiotic antibodies bound to magnetic beads. Incubated at 4℃ for 15minutes.

(5) PBS washed twice and resuscitated.

(6) The appropriate sorting procedure was selected by MACS instrument, and the negative part was taken as $CD4^+$ T cells.

Notes

1. Whole blood solutions can be added to the upper or lower layers of Ficoll, but ultimately both solutions must be delaminated clearly.

2. When separating the PBMC from the first step of centrifugation, you must not set or set a low level of braking. Otherwise, there will be stratification and confusion.

3. Antibody coated magnetic beads often have non – specific binding to dead cells and red blood cells. Therefore, the removal of dead cell red blood cells before sorting can improve the separation efficiency.

4. When the sorted cells are used for culture, attention should be paid to aseptic culture in the whole process.

实验二十一　线粒体荧光染色

扫码“学一学”

扫码“看一看”

【导言】

线粒体是大部分细胞内存在的具有双层膜结构的细胞器，其在生理及病理过程中都发挥了重要作用。线粒体不仅被称为“细胞能量工厂”，通过合成 ATP 为细胞提供能量，许多研究结果表明，线粒体还参与调节氧化还原、钙离子平衡以及细胞死亡等过程。线粒体的形态具有高度可变性，外界刺激能够诱导线粒体发生分裂或相互融合。因此，通过对线粒体形态变化的观察，可以了解不同外界刺激下线粒体功能的变化以及细胞状态的改变。

【目的】

1. 掌握线粒体荧光染色的操作步骤，并利用荧光倒置显微镜观察活细胞线粒体形态。
2. 了解线粒体荧光染色的原理。

【原理】

线粒体内膜存在的跨膜离子梯度称为线粒体膜电位（mitochondrial membrane potential, MMP）。MMP 是 ATP 合成所必需的，MMP 的减少会引起 ATP 合成下降以及线粒体内膜空间和基质 pH 改变。活细胞所含线粒体能够被特殊的荧光染料标记，这类荧光染料能够在线粒体聚集，而其聚集受线粒体内 pH 的调节。通过对线粒体荧光染料聚集多少和聚集形态的观察，能够充分了解外界刺激下线粒体的完整性以及功能的变化。

MitoTracker Red CMXRos 是一种 rosamine 衍生物，其所含有的弱巯基反应性氯甲基（thiol - reactive chloromethyl）官能团能特异性地标记具有生物活性的线粒体，是一种红色荧光染料（图 21 - 1）。MitoTracker Red CMXRos 具有良好的细胞通透性，可通过被动转运穿过细胞膜，在活细胞线粒体中聚集。弱巯基反应性氯甲基可以和线粒体内蛋白的巯基反应并共价连接，因此，使用该染料标记线粒体后，可继续使用醛类固定剂以及细胞通透去垢剂（如 Triton X - 100）等处理细胞。醛类固定剂以及细胞通透去垢剂不会影响 MitoTracker Red CMXRos 对线粒体的荧光标记效率。另一方面，由于红色荧光光谱与蓝绿色荧光光谱间的差异，MitoTracker Red CMXRos 也适用于双荧光或多荧光标记实验。

MitoTracker Red CMXRos 可溶于 DMSO，用于线粒体荧光标记实验时常配制为 1mM 的储存液。MitoTracker Red CMXRos 储存液需要避光保存于 - 20℃，保存时间长达 6 个月。MitoTracker Red CMXRos 的实际工作浓度需根据不同细胞株以及实验过程中的其他影响因素进行调整。

【试剂与器材】

1. 细胞培养基。
2. 磷酸盐缓冲液（PBS）。

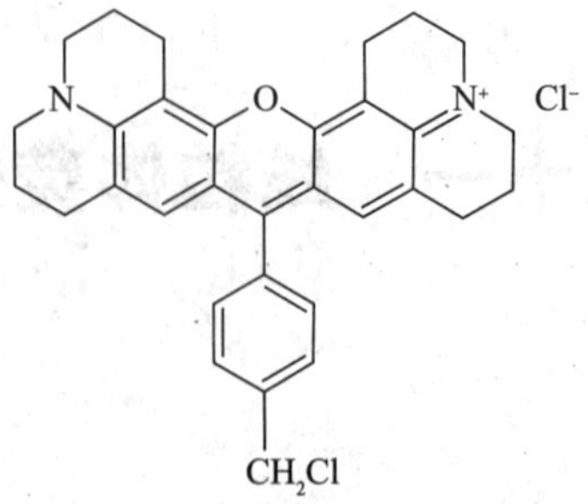

图 21-1 MitoTracker Red CMXRos 结构

注：分子式：$C_{32}H_{32}Cl_2N_2O$；分子质量：531.5236；名称：1*H*，5*H*，11*H*，15*H* - Xantheno［2，3，4 - ij：5，6，7 - i'j'］diquinolizin - 18 - ium，9 - ［4 - （chloromethyl）phenyl］；- 2，3，6，7，12，13，16，17 - octahydro - chloride；CAS 编号：167095 - 09 - 2；激发波长（E_x）：579nm；发射波长（E_m）：599nm

3. MitoTracker Red CMXRos。
4. 37% 甲醛溶液（formaldehyde solution）。
5. 0.2% 聚乙二醇辛基苯基醚（Triton X - 100 solution）。
6. 4，6 - 二脒基 - 2 - 苯基吲哚（4，6 - Diamidino - 2 - phenylindole，DAPI）。
7. 荧光 - 显微镜。

【实验步骤】

（一）细胞准备

1. 实验前一天进行细胞铺板，实验当天，细胞密度为 50% ~60% 融合。
2. 将细胞放置于 37℃，含有 5% CO_2的培养箱中培养 24 小时。

（二）线粒体荧光染料的准备

1. 利用细胞培养基，将 1mM MitoTracker Red CMXRos 储存液稀释为终浓度 100nM 用于染色。
2. 将染色液置于 37℃温浴。

（三）细胞染色

1. 移除细胞培养液，使用预热的磷酸缓冲液对细胞进行漂洗。
2. 加入步骤（二）中配制的线粒体荧光染色液。
3. 将细胞放置于 37℃，含有 5% CO_2的培养箱中孵育 20 分钟。

（四）细胞漂洗

染色孵育结束后，弃除染色液。利用预热的细胞培养液进行漂洗。

（五）细胞固定

1. 小心移除细胞培养液，加入预热的含有 3.7% 甲醛溶液的新鲜配制的培养基。
2. 将细胞放置于 37℃，含有 5% CO_2的培养箱中孵育 10 分钟。

（六）细胞漂洗

固定结束后，利用磷酸缓冲液轻柔地漂洗细胞 3 次。

（七）细胞透化

1. 将含有 0.2% Triton X - 100 的磷酸缓冲液加入细胞中，室温孵育 10 分钟。

2. 透化结束后，利用磷酸缓冲液轻柔地漂洗细胞 3 次。

（八）细胞核染色

将细胞置于含有 0.1% DAPI 的磷酸缓冲盐溶液中，室温孵育 10 分钟。

（九）荧光图像观察

利用荧光倒置显微镜，观察细胞核及线粒体形态（图 22 -2）。

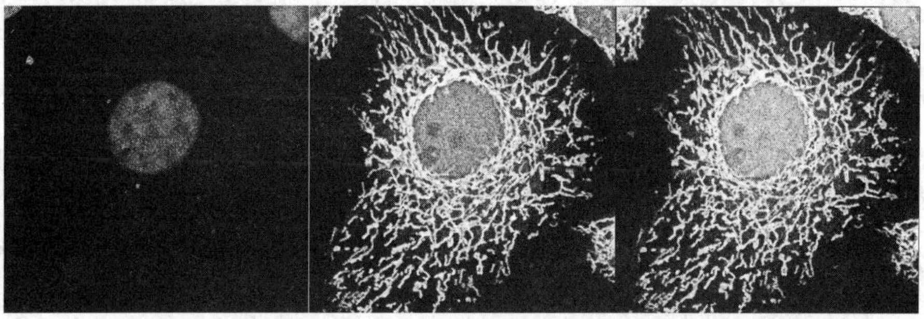

图 21 -2　HeLa 细胞的细胞核及线粒体荧光染色（中国药科大学周君提供）

【注释】

1. 实验开始前，检查细胞状态是否健康，细胞是否培养在合适的培养液中。线粒体荧光染料可在避光的状态下，置于 -20℃ 保存 6 个月。超过 6 个月的染料可能发生降解，实验前需要重新配制线粒体荧光染料储存液。

2. 线粒体荧光染料是细胞完全培养基中所含血清的潜在氧化酶，因此染色过程中应使用无血清培养基。

3. 实验过程中，需要保持细胞表面湿润，避免完全干燥。干燥后的细胞容易呈现高水平的背景荧光，影响实验结果的判断。

4. 如果实验过程中需要处理多个细胞样本，每次移除 1 ~ 2 孔的培养基，避免细胞干燥。

5. 如果线粒体染色的荧光信号过低，需要增加染料的浓度或染色时间。

Experiment 21　Fluorescence staining of mitochondria

Introduction

Mitochondria are double membrane bound organelles that play important roles in both physiological and pathological conditions. These organelles have been recognized as "powerhouse of the cell", also, a lot of evidences revealed the involvement of mitochondria in regulating redox balance, calcium balance and cell death. Mitochondria are highly dynamic in shapes that they may undergo fission and fusion with each other in response to different stresses. Therefore, it is critical to be capable to visualize mitochondrial shape changes in order to decipher the function of mitochondrial upon different stresses.

Purpose

Understand the principle of mitochondria staining using fluorescence probe. Perform the fluorescence probe - based staining of mitochondria and visualize the organelles using an inverted fluorescence microscope.

Principle

The inner membrane of the mitochondria maintains a transmembrane gradient of ions that is called mitochondrialmembrane potential (MMP). It is known that MMP is required for ATP production. Reduction of MMP leads to ATP decrease as well as pH changes within mitochondrial intermembrane space and matrix. Active mitochondria in living cells can be labeled with specific fluorescent dye that accumulate within mitochondria depending on the pH. Therefore, the changes of the fluorescent dye accumulationwillhelp to evaluate the condition of mitochondria integrity and function.

MitoTracker Red CMXRos is a rosamine - based dye that contains a mildly thiol - reactive chloromethyl moiety for mitochondria labeling (Fig. 21 - 1). This dimethyl sulfoxide (DMSO) soluble dye can passively diffuse across the plasma membrane and accumulate in active mitochondria in living cells. MitoTracker Red CMXRos is well retained after an aldehyde - based fixation and permeabilization, therefore, it is suited for double or even multiple color labeling assays.

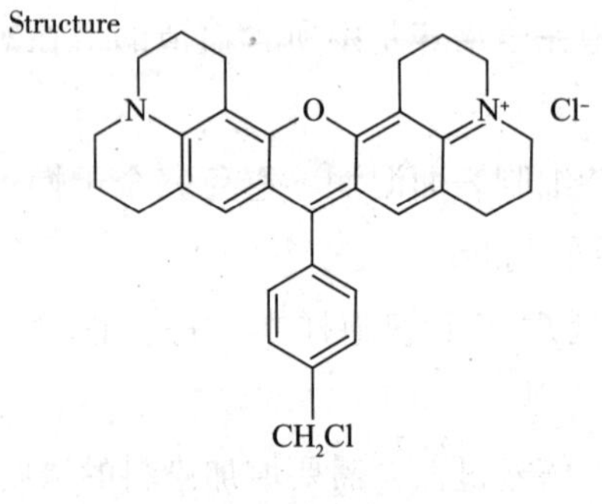

Fig. 21 - 1 Structure of MitoTracker Red CMXRos

Note: Molecular formula: $C_{32}H_{32}Cl_2N_2O$; Molecular Weight: 531.5236 Name: 1*H*, 5*H*, 11*H*, 15*H* - Xantheno [2,3,4 - ij:5,6,7 - i'j'] diquinolizin - 18 - ium, 9 - [4 - (chloromethyl) phenyl] - 2,3,6,7,12,13,16,17 - octahydro - chloride; CAS number: 167095 - 09 - 2; Excitation (E_x): 579nm Emission (E_m): 599nm

The MitoTracker Red CMXRos staining solutioncan be dissolved in high - quality DMSO to a concentration of 1mM as stock. The stock solution should be protected from light and stored at -20℃. The final working concentrations may be adjusted based on the distinct cell type or on other factors.

Reagents and apparatus

1. Cell culture Medium.

2. Phosphate - buffered saline (PBS).

3. MitoTracker Red CMXRos.

4. 37% formaldehyde solution.

5. 0. 2% Triton X – 100 solution.

6. 4,6 – Diamidino – 2 – phenylindole(DAPI).

7. Fluorescence microscope.

Procedures

1. Cells preparation

(1)Cells are seeded oncoverslips with the appropriate culture medium for a confluency of 50% – 60% the next day.

(2)Incubate the cells for the next 24hours undisturbed at 37℃ with 5% CO_2.

2. Staining solution preparation

(1)Dilute 1 mM MitoTracker Red CMXRosstock solution to the final 100nM working concentration in growth medium.

(2)Pre – warm the staining solution to 37℃.

3. Cells staining

(1)Remove the media from the cells and add gently wash the cells with pre – warmed PBS buffer.

(2)Add staining solutionprepared in step 2.

(3)Incubation cells at 37℃ with 5% $CO_2$for 20minutes.

4. Washing the cells

After staining, gently wash the cellsinpre – warmed growth medium.

5. Fixing the cells

(1)Carefully remove the medium covering the cells, and replace it with freshly prepared, pre – warmed growth medium containing 3. 7% formaldehyde.

(2)Incubate the cells at 37℃ for 15minutes.

6. Rinsing the cells

After fixation, gently rinse the cells three times with PBS buffer.

7. Permeabilization

(1)Incubate fixed cells in PBS containing 0. 2% Triton X – 100 for 10minutes at room temperature.

(2)Following permeabilization, rinse the cells in PBS buffer for three times.

8. Nucleus staining

Proceed with nucleus staining using DAPI(1 : 1000 dilution in PBS) at room temperature for 10minutes.

9. Fluorescence image visualization

Mitochondria and nucleus can be visualized under a fluorescent microscope(Fig. 21 – 2).

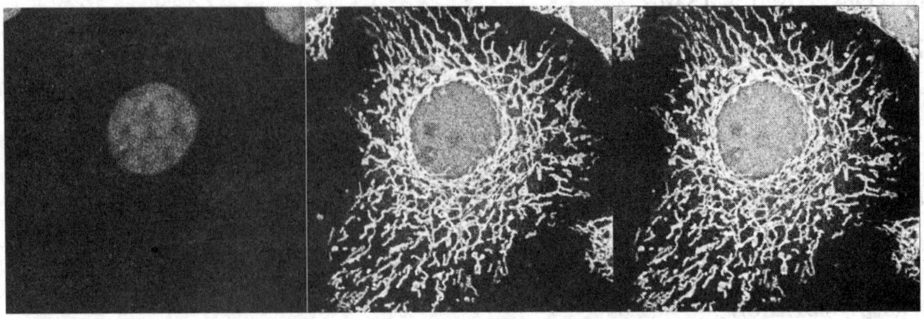

Fig. 21 – 2 Immunofluorescent analysis of HeLa celllabeling mitochondria and nucleus (provided by Jun Zhou, China Pharmaceutical University)

Notes

1. Before starting the experiment, make sure the cells are healthy and in appropriate growth medium. Double check the MitoTracker stock solution, the MitoTracker dye might degrade beyond 6months.

2. The MitoTracker dyes are potential oxidases in serum, so avoid to use complete cell culture medium with this dye.

3. Do not dry out cells during the entire process as it will lead to increased levels of background staining and difficulty in interpretation of staining results.

4. If there are multiple wells to be operated, it is best not to aspirate more than 2 wells at one time in order to reduce the possibility of the cells within the wells drying.

5. If the fluorescent signal is low, increase the MitoTracker staining concentration or incubation time.

实验二十二　动物细胞转染

扫码"学一学"

扫码"看一看"

【导言】

转染（transfection）是真核细胞主动或被动导入外源 DNA 片段而获得新的表型的过程。常规转染技术可分为两大类，即瞬时转染和稳定转染（永久转染）。前者外源 DNA/RNA 不整合到宿主染色体中，因此一个宿主细胞中可存在多个拷贝数，产生高水平的表达，但通常只持续几天，多用于启动子和其他调控元件的分析，一般需要在转染后 24 ~ 72 小时内分析结果。后者外源 DNA 可以整合到宿主染色体中，通过选择性标记筛选，得到稳定转染的细胞系。

转染技术的选择对转染结果影响也很大，许多转染方法需要优化 DNA 与转染试剂比例、细胞数量、培养及检测时间等。目前主要的转染方法包括化学和物理两大类。化学类包括：DEAE - 葡聚糖法、磷酸钙法、病毒感染法、人工脂质体法等；物理类包括：显微注射、电穿孔、基因枪等。理想的细胞转染方法应具有高转染效率、对细胞的毒性和副作用小等。本实验主要介绍细胞转染常用方法——脂质体转染的原理和操作步骤。

【目的】

1. 学习和掌握外源基因导入真核细胞的主要方法——脂质体介导的转染。
2. 了解外源基因进入的一般性方法，观测外源蛋白的表达（绿色荧光蛋白）。

【原理】

脂质体是体内和体外输送载体的工具。①中性脂质体：利用脂质膜包裹 DNA，借助脂质膜将 DNA 导入细胞膜内。②阳离子脂质体：带负电荷的 DNA 并没有预先包埋在脂质体中，而是结合到带正电荷的脂质体上，形成 DNA - 阳离子脂质体复合物，从而吸附到带负电荷的细胞膜表面，经过内吞导入细胞。此方法适用于转染悬浮或贴壁培养细胞，是目前条件下较方便的转染方法之一。

【试剂与器材】

1. 待转染细胞（293T 细胞）。
2. Lipofectamine 2000 转染试剂。
3. 质粒 DNA（0.5 ~ 5.0μg/μl 储液，EGFP 表达质粒）。
4. Opti - MEM 培养基。
5. DMEM 培养基。
6. 磷酸盐缓冲液（PBS）。
7. 胰酶/EDTA 消化液。
8. 细胞计数板。

9. 转染管。

10. 微量离心管。

【实验步骤】

1. 细胞铺板：取6孔细胞培养板，向每孔中加入2ml含有约2×10^5个细胞培养液，混匀使其均匀分布，37℃，CO_2培养箱培养至70%～90%汇合度（过夜，第二天转染）。

2. 转染试剂准备：在EP管中制备以下两个液体（转染每一个孔细胞所用的量）。

（1）A液　Opti－MEM培养基稀释2μg DNA，终量100μl。

（2）B液　Opti－MEM培养基稀释4～6μl Lipofectamine 2000，终量100μl。

将A液与B液轻弹混匀，室温静置5分钟。

3. 吸取B液加入A液中，轻弹混匀，室温静置10～15分钟。

4. 转染：将A/B混合物缓缓加入6孔板培养基中，摇匀，37℃孵育4～6小时后，换入正常培养基继续培养。

5. 瞬时转染后，可在24小时后通过显微镜观察GFP检测细胞转染效率，或24～72小时细胞长满后，提取细胞蛋白或RNA验证敲减/过表达效果。

【注释】

1. 转染前一天接种细胞于细胞培养板中，根据不同细胞类型和生长速度，保证转染时细胞汇合度达到70%～90%。

2. 准备转染复合物时，注意DNA或转染试剂要稀释在无血清、无抗生素的培养液中，整个操作最好在30分钟内完成。

3. 转染复合物直接加入含细胞培养基的细胞中，根据不同转染试剂类别，需注意是否是血清和（或）抗生素存在或不存在时均可。

4. 转染试剂用量各有不同，需提前测试不同的DNA和转染试剂的浓度和比例，以确定最佳用量。

5. 将转染后的细胞放入培养箱孵育4～6小时后，可以更换含血清的正常培养基去除复合物（也可不用）。

Experiment 22　Animal cells transfection

Introduction

Transfection is the process whereby eukaryotic cells can be transformed with either active or passive transfection of exogenous DNA fragments. Conventional transfection techniques can be divided into two categories: transient transfection and stable transfection (permanent transfection). The former foreign DNA/RNA is not integrated into the host chromosome, so multiple copy Numbers can exist in a host cell to produce high levels of expression, which usually lasts only a few days and is often used for analysis of promoters and other regulatory elements. In general, the results were analyzed within 24 – 72hours after transfection. The latter is also known as stable transfection, in which exogenous DNA can be integrated into host chromosomes and stable transfected cell lines can be obtained through selective marker screening.

The selection of transfection techniques also has a great impact on transfection results. Many transfection methods need to optimize the ratio of DNA to transfection agents, the number of cells, culture and detection time, etc. At present, the major transfection methodsinclude chemical and physical. Chemical categories include: DEAE – glucan method, calcium phosphate method, virus infection method, artificial liposome method, etc. Physics includes microinjection, electroporation, gene gun, etc. The ideal cell transfection method should have high transfection efficiency and low toxic and side effects on cells. This paper mainly introduces the principle and procedure of liposome transfection.

Purpose

To learn and master the main method of introducing foreign genes into eukaryotic cells—liposome – mediated transfection. To understand the general methods of foreign gene entry and observe the expression of foreign protein (green fluorescent protein).

Principle

Liposome is used as a carrier in vitro and in vivo. Neutral liposomes use lipid membrane to wrap DNA and introduce DNA into cell membrane. In the charged cationic liposomes, DNA is not pre – encapsulated in the liposomes, but the negatively charged DNA binds to the positively charged liposomes, forming a DNA – cationic liposome complex, which is adsorbed to the negatively charged cell membrane surface and introduced into the cells through endocytosis. This method is suitable for transfection of suspended or adherent cultured cells.

Reagents and apparatus

1. Cells to be transfected (293T cells).
2. Lipofectamine 2000 transfection reagent.
3. plasmid DNA (0.5 – 5.0μg/μl storage, EGFP expression plasmid).
4. Opti – MEM medium.
5. DMEM medium.
6. phosphate buffer (PBS).
7. trypsin /EDTA digestive fluid.
8. Hemocytometer.
9. Transfection tube.
10. Microcentrifuge tube.

Procedures

1. Cell planking: a 6 – well cell culture plate was taken, 2ml of cell culture medium containing about 2×10^5 cells was added to each well, which was mixed to make it evenly distributed. CO_2 incubator was cultured at 37℃ to 70% – 90% confluence degree (overnight, transfection on the second day).

2. Preparation of transfection agents: prepare two fluids (the amount used to transfect each pore cell) in an EP tube.

(1) Solution: Opti – MEM medium diluted 2μg DNA final volume 100μl.

(2) Solution: Opti – MEM medium diluted 4 – 6μl Lipofectamine 2000 final volume 100μl.

A and B were lightly shaken and mixed, respectively, and let stand at room temperature for 5min.

3. Absorb liquid B and add it to liquid A, lightly mix it, and let it stand at room temperature for 10 – 15min.

4. Transfection: the mixture of A/B was slowly added to the 6 – well plate culture medium, shaken well, incubated at 37℃ for 4 – 6h, and continued to be cultured in the normal culture medium.

5. After transient transfection, the transfection efficiency of cells could be detected by observing GFP under a microscope 24h after transfection, or cell protein or RNA could be extracted after 24 – 72h after cell overgrowth to verify the knockdown/overexpression effect.

Notes

1. Inoculate cells in the cell culture plate one day before transfection, and ensure that the confluence of cells reaches 70% – 90% according to different cell types and growth rates.

2. When preparing the transfection complex, it should be noted that the DNA or transfection reagent should be diluted in the culture medium without serum or antibiotics, and the whole operation should be completed within 30min.

3. The transfection complex was added directly to cells containing the cell culture medium. According to different transfection reagent categories, whether serum/antibiotic was present or not was sufficient.

4. The transfection reagent dosage varies. Different DNA and transfection reagent concentrations and ratios should be tested in advance to determine the best dosage.

5. After the transfected cells wereincubated in an incubator for 4 – 6h, normal medium containing serum could be replaced to remove the complex (or not).

实验二十三　细胞增殖检测（MTT 检测）

扫码“学一学”

扫码“看一看”

【导言】

3－(4，5－二甲基噻唑－2)－2，5－二苯基四氮唑溴盐（MTT）实验是一种检测细胞存活和生长的方法。本法利用活细胞的线粒体将 MTT 转化为甲臜晶体，从而测定线粒体的活性。对于大多数细胞群体而言，线粒体的总活性与活细胞的数量有关，因此该测定法广泛用于一些生物活性因子的活性检测、大规模的抗肿瘤药物筛选、细胞毒性实验以及肿瘤放射敏感性测定等。

【目的】

通过本实验掌握 MTT 实验的原理并熟悉实验的操作步骤。

【原理】

MTT 实验的一般目的是在相对较高的通量下（96 孔板）测量活细胞，而无须进行烦琐的细胞计数。对于大多数活细胞，线粒体活性是恒定的，因此活细胞数量的增加或减少与线粒体活性呈线性相关。活细胞线粒体的琥珀酸脱氢酶能使外源性 MTT 还原为水不溶性的蓝紫色结晶甲臜（Formazan）并沉积在细胞中，而死细胞无此功能。甲臜晶体可以被二甲基亚砜（DMSO）溶解，并可以通过酶联免疫检测仪检测其在 570nm 波长处的吸光度（*OD*）值，以测定甲臜的生成量，并以此来间接测量活细胞数量的增减。

将经过药物处理的细胞与未经过药物处理的细胞对应孔的 *OD* 值进行比较，可检测细胞对药物的敏感性。MTT 法适用于在已有细胞系以及原代细胞中检测细胞对药物的敏感性。对于分裂细胞（通常是细胞系），细胞数量的减少反映了药物对细胞生长的抑制作用，通常将药物敏感性规定为：与未处理的对照组相比，达到 50% 生长抑制作用所需的药物浓度（半数抑制浓度，IC_{50}）。对于原代细胞（非分裂细胞），药物的敏感性通过测量与对照组相比增强的细胞杀伤作用来反映，即半数致死浓度，LC_{50}。

【试剂与器材】

1. MTT 溶液（5mg/ml）：称取 0.5g MTT 溶解于 100ml 的磷酸盐缓冲液（PBS）或无酚红培养基中，用 0.22μm 滤膜过滤除菌后，避光保存。MTT 溶液最好现用现配，也可 4℃避光保存 2 周；长期保存应置于－20℃。

2. 含 10% 胎牛血清的细胞培养基、无血清的细胞培养基、PBS、0.25% 胰蛋白酶、二甲基亚砜（分析纯）。

3. 96 孔板、移液枪、细胞计数板。

【实验步骤】

1. 消化细胞并计数　用 0.25% 胰蛋白酶消化细胞，消化终止后离心收集细胞。加入含

1% ~2%胎牛血清的细胞培养基重悬细胞，行细胞计数，并调整浓度至（3 ~5）$\times 10^4$个/毫升，可根据实验目的及细胞生长本身特性进行相应调整。

2. 铺板并培养细胞 将单细胞悬液加入96孔板中（注意96孔板外围添加PBS以防培养基蒸发，且预留空白孔仅添加细胞培养基），每孔100μl。将96孔板置于37℃，5% CO_2恒温培养箱培养12 ~24小时。

3. 加药培养 用含5%胎牛血清的细胞培养基梯度稀释药品，每种药物设置6 ~8个浓度梯度。细胞贴壁后，小心吸弃96孔板中的培养基上清，每孔添加100μl不同浓度的药物，每个浓度至少设置3个复孔。空白组与阴性对照组重新添加含5%胎牛血清的细胞培养基。将96孔板置于37 ℃，5% CO_2恒温培养箱培养3 ~5天，具体时间根据实验目的和要求进行调整。

4. 加MTT 培养结束后每孔加入10μl MTT（1mg/ml），继续在37℃，5% CO_2恒温培养箱培养4小时。

5. 显色 取出96孔板，平板离心机3000转离心5分钟。吸尽培养基，每孔加入150μl DMSO，振摇平板5分钟，使结晶物充分溶解。

6. 比色 选择570nm和630nm（背景波长）测定各孔的吸光度值，记录结果并进行分析。

【注释】

1. 细胞操作需在无菌条件进行。
2. MTT有致癌性，配制时需戴好手套、做好防护措施，在无菌条件下进行。
3. 加入细胞悬液时应反复多次混匀，确保每孔接种的细胞密度完全相同。
4. 实验时应设置空白组、对照组、加药孔。
5. 加药时，吸弃一部分孔上清后立即加药，避免96孔板干燥引起细胞死亡。
6. 避免血清干扰，高浓度胎牛血清会影响试验孔的光吸收度值，因此一般选小于2%胎牛血清的细胞培养基进行。

Experiment 23 Cell proliferation assay (MTT assay)

Introduction

The 3 –(4,5 – dimethylthiazol – 2 – yl) – 2,5 diphenyl tetrazolium bromide(MTT) assay is a method for detecting cell survival and growth. This method based on the conversion of MTT into formazan crystals by living cells, which determines mitochondrial activity. Since for most cell populations the total mitochondrial activity is related to the number of viable cells, this assay is widely used for the detection of activity of some bioactive factors, large – scale anti – tumor drug screening, cytotoxicity testing, and measurement of tumor radio sensitivity etc.

Purpose

By the end of this laboratory exercise, you should be able to explain the principle MTT and describe how do to it.

Principle

The general purpose of the MTT assay is to measure viable cells in relatively high throughput (96 – well plates) without the needforelaborate cell counting. The principle of the MTT assay is that for most viable cells mitochondrial activity is constant and thereby an increase or decrease in the number of viable cells is linearly related to mitochondrial activity. The mitochondrial dehydrogenase of living cells can reduce exogenous MTT to water – insoluble blue – violet crystalline formazan and deposit in cells, whereas dead cells do not. Hyperthyroid crystals can be dissolved in dimethyl sulfoxide (DMSO), and the absorbance (*OD*) value at 570nm can be detected by an enzyme – linked immunosorbent assay to determine the amount of formazan produce and indirectly the increase or decrease of cell number.

The MTT assay is suitable for the measurement of drug sensitivity in established cell lines as well as primary cells. For dividing cells (usually cell lines) the decrease in cell number reflects cell growth inhibition and the drug sensitivity is then usually specified as the concentration of the drug that is required to achieve 50% growth inhibition as compared to the growth of the untreated control (50% inhibitory concentration, IC_{50}).

For primary (no dividing) cells, drug sensitivity is measured as enhanced cell kill of treated cells as compared to the loss of cells already commonly seen in untreated cells (50% lethal concentration, LC_{50}).

Reagents and apparatus

1. MTT solution (5mg/ml): Weigh 0. 5g of MTT dissolved in 100ml of phosphate buffer (PBS) or phenol red free medium. Solution must be filter sterilized with 0. 22μm filter, and store in the dark place. Reconstituted MTT solution is stable for at least 6 months when stored at – 20℃. Storage at 4℃ for more than four days will result in decomposition and will yield erroneous results.

2. Cell culture medium containing 10% fetal bovine serum, serum – free cell culture medium, PBS, 0. 25% trypsin and dimethyl sulfoxide (analytical grade).

3. 96 – well plate, pipette, cell counting plate.

Procedures

1. Digest the cells and count: digest the cells with 0. 25% trypsin then centrifuge the cells after centrifugation. The cells wereresuspended in a cell culture medium containing 1% – 2% fetal bovine serum, and the cells were counted and adjusted to a concentration of $(3-5)\times 10^4$/ml (the corresponding cell density can be adjusted according to the purpose of the experiment and the characteristics of the cell growth itself).

2. Plate and culture the cells: Add the single cell suspension into the 96 – well plate (note that PBS is added to the periphery of the 96 – well plate to prevent evaporation of the medium, and blank cells are reserved for cell culture only), 100μl per well. Thenincubate it for 12 – 24h in a 37℃, 5%

CO_2 incubator.

3. Dosing culture: The drug was diluted with a cell culture medium containing 5% fetal bovine serum, and each drug was set to 6 – 8 concentration gradients. For adherent cells, carefully remove the medium and replace it with 100μl of different concentrations of the drug. Each concentration of drug must be replicate at least 3times. The blank medium and the negative control group were replaced to the cell culture medium containing 5% fetal bovine serum. Then incubate it for 3 – 5days in 37℃, 5% CO_2 incubator. Specific time can be adjusted according to the purpose and requirements of the experiment.

4. Add MTT: Add 10μl of MTT(1 mg/ml) to each well thenincubate it for 4h at 37℃ in 5% CO_2 incubator.

5. Color development: The 96 – well plate was taken out and centrifuged at 3000rpm for 5minutes in a plate centrifuge. The medium was carefully removed, 150μl of DMSO was added to each well, and the plate was shaken for 5minutes to dissolve the crystals sufficiently.

6. Colorimetric: Select 570nm and 630nm(background wavelength) to determine the absorbance of each well, record the results and analyze.

Notes

1. Cell manipulation should be performed under aseptic conditions.

2. MTT is carcinogenic. Aseptically add it in an amount equal culture volume.

3. When adding cell suspense, mix repeatedly and make sure that the density of cells inoculated per well is exactly the same.

4. Blank group, control group and dosing hole should be set during the experiment.

5. When adding the drug, a part of the well should be discarded and immediately added to avoid cell death caused by drying of the 96 – well plate.

6. Avoid serum interference. High concentrations of fetal bovine serum affect the absorbent. Therefore, a cell culture medium must be containing less than 2% fetal bovine serum.

实验二十四　细胞凋亡的诱导与检测

扫码“学一学”

扫码“看一看”

【导言】

细胞凋亡（apoptosis）一词来源于古希腊语，意思是花瓣或树叶凋落，意味着生命走到了尽头，细胞到了一定时期会像树叶那样自然死亡。

凋亡是细胞在一定生埋或病理条件下遵守自身程序的主动死亡过程。凋亡时细胞皱缩，表面微绒毛消失，染色质凝集并呈新月形或块状靠近核膜边缘，继而核裂解，由细胞膜包裹着核碎片或其他细胞器形成小球状凋亡小体突出于细胞表面，最后凋亡小体脱落被吞噬细胞或邻周细胞吞噬。凋亡过程中溶酶体及细胞膜保持完整，不引起炎症反应。细胞凋亡在个体正常发育、稳态维持、免疫耐受形成、肿瘤监控和抵御各种外界因素干扰等方面都起着关键性的作用。

【目的】

通过本实验学习细胞凋亡的概念，并掌握检测细胞凋亡的基本方法。

【原理】

（一）细胞凋亡的概念及机制

细胞凋亡指为维持内环境稳定，由基因控制的细胞自主的有序的死亡，又称细胞程序性死亡。细胞凋亡与细胞坏死不同，细胞凋亡是一个主动过程，涉及基因的激活、表达以及调控等一系列作用，它并不是病理条件下自体损伤的一种现象，而是为更好地适应生存环境主动进行的死亡过程。

（二）细胞凋亡的检测方法

常用的细胞凋亡的检测方法包括：细胞凋亡的形态学检测、磷脂酰丝氨酸外翻分析（Annexin V 法）、线粒体膜势能（$\Delta\Psi mt$）检测、DNA 片段化检测、TUNEL 法、Caspase－3 活性的检测、WB 检测和凋亡相关蛋白的表达及细胞定位分析等。

细胞凋亡的形态学检测表现如下。①未染色细胞：凋亡细胞的体积变小、变形，细胞膜完整但出现发泡现象，细胞凋亡晚期可见凋亡小体。贴壁细胞出现皱缩、变圆、脱落。②染色细胞：常用姬姆萨染色、瑞氏染色等。凋亡细胞呈现染色质浓缩、边缘化，核膜破裂，染色质分割成块状和出现凋亡小体等典型的凋亡形态。

【试剂】

1. HeLa 细胞。
2. 含 10% 小牛血清的 DMEM 高糖培养基。
3. 甲醇固定液。
4. PBS。

5. VP－16。

6. 吖啶橙染液。

7. 瑞氏染液。

【实验步骤】

（一）细胞培养和凋亡药物诱导处理

1. 在细胞培养室中复苏和传代培养 HeLa 细胞，使细胞处于对数生长期。

2. 实验前 2 天，将盖玻片放入 35mm 细菌培养皿，并将细胞种植在培养皿上，分为实验组和对照组。

3. 实验前 1 天，使细胞的底部汇合度为 60%～70%，加入 VP－16 至终浓度为 0.1mmol/L，在 37℃培养箱中继续孵育 24 小时。对照组加入等量的 DMSO。

（二）倒置显微镜下观察

培养完成后，从培养箱中将 35mm 培养皿取出，在倒置显微镜下进行观察，拍照记录结果。

（三）瑞氏染色后倒置显微镜观察

1. 固定 将实验组和对照组培养皿中的培养液吸去，用 PBS 清洗 1～2 遍，加入适量的预冷甲醇，冰箱中 0℃以上低温固定 10 分钟，之后取出培养皿，吸去甲醇，用 PBS 清洗 2～3 遍。

2. 染色 向已取出盖玻片的培养皿中加入适量的瑞氏染液（稍没过皿底即可），染色 10～15 分钟，之后吸去染液，用缓冲液清洗，洗去浮色。之后倒掉缓冲液，静置干燥。

3. 观察 将干燥的培养皿放在倒置显微镜下观察，拍照记录结果。正常细胞核染成深蓝色或蓝紫色，色泽均一，凋亡细胞可见核固缩、破碎，染色变深，细胞膜皱缩、出泡，出现凋亡小体。凋亡细胞至少要选取 5 个不同视野拍照，用于之后计算凋亡率。

（四）计算凋亡率

计数视野中的细胞总数和凋亡细胞数，制作成表，并按照如下公式计算凋亡率：

凋亡率＝凋亡细胞数/细胞总数

Experiment 24 Induction and detection of apoptosis

Introduction

Apoptosis comes from the ancient Greek, which means that the petals or leaves fall, meaning that life has come to an end, and cells will die naturally like leaves in a certain period.

Apoptosis is the active death process of cells following their procedures under certain physiological or pathological conditions. During apoptosis, the cells shrink, the surface microvilli disappear, the chromatin aggregates, and form a crescent or block near the edge of the nuclear membrane, and then the nuclear is lysed. The nuclear membrane or other organelles are wrapped by the cell membrane to form a small spherical apoptotic body. On the cell surface, the apoptotic bodies finally shed and were phagocytosed by phagocytic cells or neighboring cells. The lysosome and cell membrane remained intact during the apoptotic process without causing an inflammatory response. Apoptosis plays a crucial role in the individual's healthy development, homeostasis maintenance, immune tolerance formation, tumor monitoring, and resistance to various external factors.

Purpose

Through this experiment, we can learn the concept of apoptosis and master the basic methods of detecting apoptosis.

Principle

1. Definition and mechanism of apoptosis

Apoptosis refers to the orderly death of cells that are controlled by genes to maintain homeostasis, also known as programmed cell death. Apoptosis is different from cell necrosis. Apoptosis is not a passive process, but an active process. It involves the activation, expression, and regulation of a series of genes. It is not a kind of autologous injury under pathological conditionsphenomenon, but a death process that is actively pursued to better adapt to the living environment.

2. Detection method of apoptosis

Common methods for detecting apoptosisinclude morphological detection of apoptosis, phosphatidylserine valgus analysis (Annexin V method), mitochondrial membrane potential energy ($\Delta\Psi$mt) detection, DNA fragmentation detection, TUNEL method, Caspase - 3 Activity detection, WB detection, and expression of apoptosis - related proteins and cell localization analysis.

The results of morphological detection of apoptosis are as fonows: ①Unstained cells: The volume of apoptotic cells becomes smaller and deformed, the cell membrane is intact but foaming occurs, and apoptotic bodies are seen in the late stage of apoptosis. Adherent cells appear shrunk, round, and fall off. ②Stained cells: Giemsa staining, Wright's staining, etc. are commonly used. The apoptotic cells are characterized by chromatin condensation, marginalization, nuclear membrane rupture, chromatin segmentation into massive and apoptotic bodies.

Reagents and apparatus

1. HeLa cells.
2. DMEM high glucose medium containing 10% calf serum.
3. methanol fixative.
4. PBS.
5. VP - 16.
6. acridine orange dye solution.
7. Wright's dye solution.

Procedures

1. Cell culture and apoptosis drug induction treatment

(1) Resuscitation and subculture of HeLa cells in a cell culture chamber to place the cells in a

logarithmic growth phase.

(2)2g before the experiment, thecoverslips were placed in a 35mm bacterial culture dish, and the cells were planted on a petri dish and divided into an experimental group and a control group.

(3)1d before the experiment, the bottom confluence of the cells was about 60% -70%, VP-16 was added to a final concentration of 0.1mmol/L, andincubation was continued for 24hours in a 37℃ incubator. The control group was added an equal amount of DMSO.

2. Observed under an inverted microscope

After the completion of the culture, a 35mm culture dish was taken out from theincubator, observed under an inverted microscope, and photographed to record the results.

3. Inverted microscope observation after Wright's staining

(1) Fixation: Absorb the culture solution in the experimental group and the control culture dish, wash it with PBS for 1-2times, add appropriate amount of pre-cooled methanol, fix it in the refrigerator for 10min at low temperature, then take out the culture dish and suck off the methanol. Wash 2-3 times with PBS.

(2) Dyeing: Add appropriate amount of Wright's dye solution to the Petri dish with the cover slip removed (slightly without the bottom of the dish), stain for 10-15min, then remove the dye solution, wash with buffer, wash and float color. After that, the buffer was poured off and allowed to stand to dry.

(3) Observation: The dried culture dish was placed under an inverted microscope, and the results were photographed. The normal cell nucleus is dyed dark blue or blue-violet, and the color is uniform. The apoptotic cells can be seen to have nuclear condensation, breakage, deep staining, cell membrane shrinkage, foaming, and apoptotic bodies. Apoptotic cells should be photographed with at least 5 different fields of view for later calculation of apoptotic rate.

4. Calculate the rate of apoptosis

Photographs of apoptotic cells under the microscope were taken, the total number of cells in the field of view and the number of apoptotic cells were counted, and a table was prepared, and the apoptosis rate was calculated according to the following formula:

Apoptosis rate = number of apoptotic cells/total number of cells

实验二十五　小胶质细胞激活的诱导与检测

扫码“学一学”

扫码“看一看”

【导言】

小胶质细胞（microglial）是中枢神经系统固有的免疫细胞，广泛分布于大脑和脊髓中。静息状态及适度活化的小胶质细胞可发挥重要的免疫防御、免疫监视作用，维持神经系统内环境稳定。过度活化的小胶质细胞则介导神经炎症，释放包括 NO，TNF－α 及活性氧簇在内的炎症介质，导致神经损伤。BV－2 细胞系为 Blasi 等在 1990 年应用携带癌基因 *v－raf/v－myc* 的反转录病毒 J2 感染原代培养的小鼠小胶质细胞而获得的永生细胞系，具备原代培养的小胶质细胞的形态学、表型以及各项功能特点，经诱导激活后释放多种炎症介质。其中，NO 在体内由一氧化氮合酶（nitric oxide synthase，NOS）在多种辅助因子的参与下催化左旋精氨酸酶（L－Arg）胍基末端的氮原子与氧分子反应而生成，可介导并调节机体的多种生理和病理过程，在神经系统、心血管系统、炎症和免疫过程中都发挥重要作用。

【目的】

通过本实验了解小胶质细胞激活的诱导与检测（NO 检测）的原理与操作步骤。

【原理】

脂多糖（Lipopolysaccharide，LPS）是革兰阴性菌细胞外壁的组成成分，Toll 样受体 4（Toll－like receptors 4，TLR4）是其主要的识别受体。TLR4 激活后通过一系列级联反应，最终使核因子 κB（Nuclear factor kappa B，NF－κB）活化。NF－κB 被激活后启动免疫和炎症相关基因的转录，引起众多基因的互相交叉表达，导致炎症过程升级。实验中常用 LPS 刺激诱导 BV－2 细胞过度活化，促进 NO 的分泌。NO 释放后极不稳定，在体内或水中易氧化生成 NO_2^-，在酸性条件下溶液中的 NO_2^- 可与对氨基苯磺酸发生重氮化，再与 *N*－（1－萘基）－乙二胺偶联，生成紫红色产物。在 540nm 处比色，吸光度值的大小与 NO_2^- 的含量成正比，从而间接反映 NO 的生成量。

【试剂】

1. BV－2 细胞。
2. LPS 溶液。
3. DMEM 培养基。
4. FBS。
5. $NaNO_2$。
6. Griess Reagent Ⅰ：对氨基苯磺酸 0. 5g、稀醋酸（10%）150ml。
7. Griess Reagent Ⅱ：α 萘胺 0. 1g、蒸馏水 20ml、稀醋酸（10%）150ml。

【实验步骤】

（一）BV－2 细胞贴壁培养

1. 配制含有10% FBS 的 DMEM 培养基。

2. 使用上述培养基将 BV－2 细胞调整为 1×10^6 个/毫升，接种于6孔细胞培养板中，每孔加入1ml 细胞悬浮液，37℃，5% CO_2 恒温培养箱中过夜培养。

（二）LPS 刺激 BV－2 细胞

1. 配制含有1% FBS 的 DMEM 培养基。

2. 使用上述培养基配制浓度分别为1μg/ml、2μg/ml 的 LPS。

3. 弃去6孔细胞培养板中的液体，分别加入1μg/ml、2μg/ml LPS，37℃，5% CO_2 恒温培养箱中培养12小时。

（三）Griess 法测定上清液 NO

1. 使用含有1% FBS 的 DMEM 培养基稀释 $NaNO_2$ 标准品至浓度分别为0μM、1μM、2μM、5μM、10μM、20μM、40μM、60μM、100μM。

2. 按50微升/孔，在96孔板中加入标准品及样品。

3. 取出 Griess Reagent Ⅰ和Ⅱ，使其恢复室温。

4. 按50微升/孔，在各孔中加入室温 Griess Reagent Ⅰ。

5. 按50微升/孔，在各孔中加入室温 Griess Reagent Ⅱ。

6. 540nm 测定吸光度。

【注释】

（一）细胞铺板密度

进行6孔细胞板铺板时，保证每个孔里细胞密度均一且数量较多。细胞数量和状态会直接影响 NO 产生量，最终影响实验结果。

（二）Griess 试剂使用注意事项

1. Griess 试剂需4℃ 避光保存，半年有效。

2. Griess 试剂对人体有害，实验时需戴一次性手套进行操作，并注意有效防护以避免直接接触人体或吸入体内。

3. 如果保存不当导致溶液变色或沉淀，说明该溶液失效。

Experiment 25 Induction and detection of microglia activation

Introduction

Microglials are innate immune cells of the central nervous system and are widely distributed in the brain and spinal cord. The resting state and moderately activated microglia can play an important role in immune defense and immune surveillance, and maintain the stability of the nervous system. Excessively activated microglia cause neuroinflammation, releasing inflammatory mediators including NO, TNF – α and reactive oxygen species, leading to nerve damage. The BV – 2 cell line is an immortalized cell line obtained by Blasi et al. in 1990 using a retrovirus J2 carrying the oncogene *v – raf/v – myc* to infect primary cultured mouse microglia. This cell line has the morphology, phenotype and various functional characteristics of primary cultured microglia, and releases various inflammatory mediators after induction and activation. Among them, NO is produced by nitric oxide synthase (NOS) in the body, which is catalyzed by a variety of cofactors to catalyze the reaction of the nitrogen atom at the end of the arginine arginase (L – Arg) thiol group with oxygen molecules. It can mediate and regulate a variety of physiological and pathological processes in the body, playing an important role in the nervous system, cardiovascular system, inflammation and immune processes.

Purpose

This experiment is to understand the principle and operation steps of induction and detection of microglia activation (NO detection).

Principle

Lipopolysaccharide (LPS) is a component of the outer wall of Gram – negative bacteria, and Toll – like receptors 4 (TLR4) is its main recognition receptor. After activation of TLR4, nuclear factor kappa B (NF – κB) is finally activated by a series of cascade reactions. When NF – κB is activated, transcription of immune and inflammation – related genes is initiated, causing cross – expression of many genes, leading to an escalation of the inflammatory process. In the experiment, LPS stimulation was used to induce excessive activation of BV – 2 cells and promote NO secretion. NO is extremely unstable after release, and is easily oxidized to form NO_2^- in the body or in water. Under acidic conditions, NO_2^- in the solution can be diazotized with p – aminobenzenesulfonic acid and coupled with *N* – (1 – naphthyl) – ethylenediamine to form a purplish red product. at 540nm, the absorbance value is proportional to the amount of NO_2^-, which indirectly reflects the amount of NO produced.

Reagents and apparatus

1. BV – 2 cells.
2. Lipopolysaccharides (LPS).
3. DMEM medium.
4. FBS.
5. $NaNO_2$.
6. Griess Reagent Ⅰ: 0.5g 4 – aminobenzenesulfonic acid, 150ml dilute acetic acid (10%).
7. Griess Reagent Ⅱ: 0.1g naphthylethylene – diamine, 20ml distilled water, 150ml dilute acetic acid (10%).

Procedures

1. Adherent BV – 2 cell culture

(1) Configuration of DMEM medium with 10% FBS.

(2) BV – 2 cells were preincubated in 6 – well plates at a density of 1×10^6 cells per well

overnight at 37℃ in a humidified incubator under 5% CO_2.

2. LPS - stimulated BV - 2 cells

(1) Configurate DMEM medium with 1% FBS.

(2) Use the above medium to configurate LPS with a concentration of 1μg/ml, 2μg/ml, respectively.

(3) Remove the medium and add 1μg/ml, 2μg/ml LPS to 6 - well plates, then place in a humidified incubator at 37℃ under 5% CO_2.

3. Determination of Supernatant nitric oxide (NO) by the Griess assay

(1) Dilute the $NaNO_2$ standard in DMEM medium containing 1% FBS to a concentration of 0μM, 1μM, 2μM, 5μM, 10μM, 20μM, 40μM, 60μM, 100μM.

(2) Add standard and sample to 96 - well plate at 50μl/well.

(3) Place Griess reagent Ⅰ and Ⅱ to room temperature.

(4) Add 50μl Griess reagent Ⅰ per well.

(5) Add 50μl Griess reagent Ⅱ per well.

(6) Determine the absorbance at 540nm using a microplate reader.

Notes

1. Cell plating density

When plating 6 - well cell plates, ensure that cell density and cell number are adequate in each well. The number and state of cells directly affect the production of NO, which ultimately impact on the experimental results.

2. Cautions in the use of Griess Reagent

(1) Store at 4℃ away from light. Stable for 6months from the date of shipment.

(2) Griess reagents are harmful to humans. Wear disposable gloves for experiments and avoid direct contact with the body or inhalation.

(3) If improper preservation leads to discoloration or precipitation of the solution, the solution will be invalid.

附　　录

一、常用核酸、蛋白质换算数据

1. 重量换算

$1\mu g = 10^{-6}g$

$1ng = 10^{-9}g$

$1pg = 10^{-12}g$

$1fg = 10^{-15}g$

2. 分光光度换算

$1A_{260}$ 双链 DNA = 50μg/ml

$1A_{260}$ 单链 DNA = 33μg/ml

$1A_{260}$ 单链 RNA = 40μg/ml

3. DNA 摩尔换算

1μg 1000bp DNA = 1.52pmol（末端 3.03pmol）

1μg pBR322 = 0.36pmol

1pmol 1000bp DNA = 0.66μg

1pmol pBR322 = 2.8μg

1kb 双链 DNA（钠盐）= 6.6×10^5 Da

1kb 单链 DNA（钠盐）= 3.3×10^5 Da

1kb 单链 RNA（钠盐）= 3.4×10^5 Da

脱氧核糖核苷的平均分子质量 = 126.7Da

4. 蛋白摩尔换算

100pmol 分子质量 100 000 蛋白质 = 10μg

100pmol 分子质量 50 000 蛋白质 = 5μg

100pmol 分子质量 10 000 蛋白质 = 1μg

氨基酸的平均分子质量 = 126.7Da

5. 蛋白质/分子质量换算

1Kb DNA = 333 个氨基酸编码容量 = 3.7×10^4 MW 蛋白质

10 000MW 蛋白质 = 270bp DNA

30 000MW 蛋白质 = 8 100bp DNA

50 000MW 蛋白质 = 1.35kb DNA

100 000MW 蛋白质 = 2.70kb DNA

二、合成引物的参数

1 *OD* 引物干粉约为 33μg；碱基的平均分子质量为 324.5。引物的分子质量和摩尔数的计算公式如下：

引物的分子质量 = 碱基数 × 碱基的平均分子质量

引物的摩尔数 = 质量数/引物分子质量

举例：一管标注为 2 *OD* 的 20 个碱基的引物，分子质量 = 20 × 324.5 = 6490，质量数 = 2 × 33 = 66μg，摩尔数 = 66/6490 = 0.010μmol = 10nmol，若要溶解 10μM（10pmol/μl）的溶液，只需加 1ml ddH_2O 充分溶解即可。此外，真空干燥的 DNA 呈干膜状或粉末状在离心管底部，开启瓶盖前先离心，然后再加入足量的水充分振荡溶解。

三、常用培养基

1. LB（Luria - Bertani）培养基 1% Tryptone，0.5% Yeast Extract，1% NaCl，pH 7.0。

2. SOB 培养基 2% Tryptone，0.5% Yeast Extract，0.05% NaCl，2.5mM KCl，10mM $MgCl_2$，pH 7.0。

3. SOC 培养基 除含有 20mM 葡萄糖外，其余组分与 SOB 相同。

4. 2 × YT 培养基 1.6% Tryptone，1% Yeast Extract，0.5% NaCl，pH 7.0。

四、常用抗生素

抗生素	贮存液		工作浓度	
	浓度（mg/ml）	保存条件（℃）	严紧型质粒（μg/ml）	松弛型质粒（μg/ml）
氨苄西林（Amp）	51（溶于水）	-20	20	60
羧苄西林（Carbenicillin）	50（溶于水）	-20	20	60
氯霉素（Cm）	34（溶于乙醇）	-20	25	170
卡那霉素（Kan）	10（溶于水）	-20	10	50
链霉素（Sm）	10（溶于水）	-20	10	50
四环素（Tet）	5（溶于乙醇）	-20	10	50

五、琼脂糖凝胶浓度与线性 DNA 分辨范围

琼脂糖凝胶浓度（%）	线形 DNA 的最佳分辨范围（bp）
0.5	1000 ~ 30 000
0.7	800 ~ 12 000
1.0	500 ~ 10 000
1.2	400 ~ 7000
1.5	200 ~ 3000
2.0	50 ~ 2000

六、遗传密码

密码子的第一位（5′端）	密码子的第二位				密码子的第三位（3′端）
	U	C	A	G	
U	UUU Phe	UCU Ser	UAU Tyr	UGU Cys	U
	UUC Phe	UCC Ser	UAC Tyr	UGC Cys	C
	UUA Leu	UCA Ser	UAA stop	UGA stop	A
	UUG Leu	UCG Ser	UAG stop	UGG Trp	G
C	CUU Leu	CCU Pro	CAU His	CGU Arg	U
	CUC Leu	CCC Pro	CAC His	CGC Arg	C
	CUA Leu	CCA Pro	CAA Gln	CGA Arg	A
	CUG Leu	CCG Pro	CAG Gln	CGG Arg	G
A	AUU Ile	ACU Thr	AAU Asn	AGU Ser	U
	AUC Ile	ACC Thr	AAC Asn	AGC Ser	C
	AUA Ile	ACA Thr	AAA Lys	AGA Arg	A
	AUG Met	ACG Thr	AAG Lys	AGG Arg	G
G	GUU Val	GCU Ala	GAU Asp	GGU Gly	U
	GUC Val	GCC Ala	GAC Asp	GGC Gly	C
	GUA Val	GCA Ala	GAA Glu	GGA Gly	A
	GUG Val	GCG Ala	GAG Glu	GGG Gly	G

注：密码子由5′→3′方向阅读

七、常用大肠埃希菌（K－12株来源）的基因型

品系	基因型
BL21（DE3）	F^-，ompT，$hsdS_B$（$r_B^- m_B^-$），gal（1 c I 857，ind1，Sam7，nin5，lacUV5－T7gene1），dcm（DE3）（B株来源）
BMH71－18mutS	Δ（lac－proAB），supE，thi，mutS215：Tn10（tet^r）/F′（traD36，$proAB^+$，$lacI^q$，lacZΔM15）
C600	supE44，hsdR，thi－1，thr－1，leuB6，lacY1，tonA21
DH5a	supE44，ΔlacU169（φ80 lacZΔM15），hsdR17，recA1，endA1，gyrA96，thi－1，relA1
HB101	supE44，hsdS20（r_B^- m_B^-），recA13，ara－14，proA2，lacY1，galK2，rpsL20，xyl－5，mtl－1，leuB6，thi－1
JM105	endA1，supE，sbcB15，thi，rpsL，Δ（lac－proAB）/F′［traD36，$proAB^+$，$lacI^q$，lacZΔM15］
JM106	F^-，endA1，gyrA96，thi，hsdR17，supE44，relA1，Δ（lac－proAB）
JM107	endA1，gyrA96，thi，hsdR17，supE44，relA1，Δ（lac－proAB）/F′［traD36，$proAB^+$，$lacI^q$，lacZΔM15］
JM108	F^-，recA1，endA1，gyrA96，thi，hsdR17，supE44，relA1，Δ（lac－proAB）
JM109	recA1，endA1，gyrA96，thi，hsdR17，supE44，relA1，Δ（lac－proAB）/F′［traD36，$proAB^+$，$lacI^q$，lacZΔM15］
JM110	dam，dcm，supE44，hsdR17，thi，leu，rpsL1，lacY，galK，galT，ara，tonA，thr，tsx，Δ（lac－proAB）/F′［traD36，$proAB^+$，$lacI^q$，lacZΔM15］

续表

品系	基因型
K802	F^-，lacY1 or Δ（lac I^q – Y）6（lac – 3），supE44，galK2，galT22，mcrA，rfbD，metB1，mcrB1，hsdR2
K803	F^-，lacY1 or Δ（lac I^q – Y）6（lac – 3），supE44，galK2，galT22，mcrA，rfbD，metB1，mcrB1，hsdR3
LE392	supE44，supF58，hsdR514，galK2，galT22，metB1，trpR55，lacY1
RR1	supE44，hsdS20，ara – 14，proA2，lacY1，galK2，rpsL20，xyl – 5，mtl – 1
TAP90	supE44，supF58，hsdR，pro，leuB，thi – 1，rpsL，lacY1，tonA1，recD1903：mini – tet
TG1	supE，hsdΔ5，thi，Δ（lac – proAB）/F′［traD36，$proAB^+$，lac I^q，lacDΔM15］
TG2	supE，hsdΔ5，thi，Δ（lac – proAB）Δ（srl – recA）306：Tn10（tet^r）/F′［traD36，$proAB^+$，lac I^q，lacZΔM15］
TH2	supE44，hsdS20（$r_B^- m_B^-$），recA13，ara – 14，proA2，lacY1，galK2，rpsL20，xyl – 5，mtl – 1，thi – 1，trpR624
TOP10F′	F′*pro*AB，*lac*I q，*lac*Z_M15，Tn10（Tet R）} *mcr*A，_（*mrr* – *hsd*RMS – *mcr*BC），_80*lac*Z_M15，_ *lac*X74，*deo*R，*rec*A1，*ara*D139，_（*ara* – *leu*）7697，*gal*U，*gal*K，*rps*L（Str R），*end*A1，*nup*Gλ^-
Y – 1089	F^-，ΔlacU169，lon – 100，araD139，rpsL，hflA150［chr：Tn10/pMC9］
Y – 1090	F^-，ΔlacU169，lon – 100，araD139，rpsL，supF，mcrA，trpC22：Tn10/pMC9

野生的大肠埃希菌具有大约4700kbp的DNA分子，上面有约2800个基因。当这些基因中有突变发生时，基因产物随之变化，并且有可能给大肠埃希菌带来可以观察到的变化。这种能够观察到的特征称为大肠埃希菌的表现型（phenotype），而基因的构成称为大肠埃希菌的基因型（genotype）。

基因型表示方法是用基因产物或其作用的名称的3个斜体字母表示，如DNA Adenine methylase→*dam*。如果不同的基因变异导致相同的作用结果，则加上一个大写字母，如Recombination→*recA*，*recB*，*recC*。

某个基因或某区域缺失时，在基因型前面加上“Δ”表示，如从*lac*到*proAB*基因缺失时表示为Δ（*lac* – *proAB*）。

主要的基因型说明如下。

1. 停止密码回复

（1）*supE*（suppressor）　变异时，即使存在停止密码子UAG，此处也会插入谷氨酰胺（Glutamine），从而使蛋白质继续合成。

（2）*supF*（suppressor）　变异时，即使存在停止密码子UAG，此处也会插入酪氨酸（Tyrosine），从而使蛋白质继续合成。

2. 重组机能缺失

（1）*recA*（recombination）　相同的DNA重组活性缺失。由于导入DNA与宿主DNA的重组受到阻碍，可以保持插入DNA的稳定性。与*rec*A13宿主菌相比，*rec*A1宿主菌能够使插入DNA稳定存在。

（2）*recB*，*C*（recombination）　内切核酸酶V变异。抑制重组并能影响放射线损伤的修复。

（3）*traD*（transmissibility）　在质粒F因子内部，与大肠埃希菌的结合有关。*traD*变异后，F因子自身的传递能力显著下降。

3. 对插入DNA具有直接作用因子的缺失

（1）*dam*（DNA adenine methylase）　宿主菌来源的腺嘌呤甲基化酶（G^mATC）缺失。

（2）*dcm*（DNA cytsine methylase）　宿主菌来源的胞嘧啶甲基化酶（C^mCWGG）缺失。

（3）*hsdR*（host specificity defective）　宿主菌来源的Ⅰ型限制酶*EcoK*（或*EcoB*）的切断部位蛋白变异。转化的外源DNA不被限制酶*EcoK*切断，可以进行克隆。

（4）*hsdM*（host specificity defective）　宿主菌来源的Ⅰ型限制酶*EcoK*（或*EcoB*）的甲基化酶部位蛋白变异。

（5）*hsdS*（host specificity defective）　宿主菌来源的Ⅰ型限制酶*EcoK*（或*EcoB*）的识别部位蛋白变异。转化的外源DNA不被限制酶*EcoK*切断，可以进行克隆。

（6）*endA*（endonuclease）　宿主菌来源的非特异性内切核酸酶Ⅰ活性缺失，可以提高纯化的质粒DNA的质量。

4. 宿主蛋白酶缺失

（1）*lon*（long form）　分解异型蛋白质的ATP依赖型蛋白分解酶活性缺失。大肠埃希菌B株原来就为lon^-。可抑制其表达的融合蛋白质的分解。

（2）*omp*（outer membrane protein）　膜结合性蛋白分解酶活性缺失。抑制表达的融合蛋白质分解。

5. 有利于选择转化子基因

（1）*lac* I^q（lactose）　乳糖操纵子中控制β-半乳糖苷酶表达的调节蛋白基因变异。由于*lac* I^q变异，表达调节蛋白过量形成，因此从乳糖启动子开始的转录过程完全受到抑制。

（2）*lacZ*（lactose）　β-半乳糖苷酶活性缺失。

（3）*lacZ*ΔM15　β-半乳糖苷酶蛋白的ω-fragment得到表达。当与多数载体质粒中所具有的α-fragment共同存在时，可使β-半乳糖苷酶的活性恢复（α-互补性）。在含有X-gal的平板培养基上，可以通过蓝白菌落的差别选择重组体。

（4）*deoR*　变异时，可以选择性地改善大分子DNA的转化。

参考文献

[1] Sambrook J., Russell D W. Molecular Cloning: A laboratory manual [M]. 3rd ed. Cold Spring Harbor Laboratory Press. 2001.

[2] Turner P. C., McLennan A. G. Instant Notes in molecular Biology [M]. Bios Scientific Publishers Limited. 1999.

[3] Protocols and Applications Guide [M]. 3rd ed. Promega Corporatiooon. 1996.

[4] Li Y., Zhao Y. Practical Protocols in Molecular Biology [M]. Science Press. New York. 1995.

[5] Zyskind J W, Bernstein S I. Recombinant DNA Laboratory Manual [M]. Academic Press, Inc. 1989.

[6] Wilson M., DeRisi J., Kristensen H. H., et al. Exploring drug - induced alterations in gene expression in mycobacterium tuberculosis by microarray hybridization [J]. Proc Natl Acad Sci USA, 1999, 96 (22): 12833 - 12838.

[7] Yang G. P., Ross D. T., Kuang W. W., et al. Combining SSH and cDNA microarrays for rapid identification of differentially expressed genes [J]. Nucleic Acids Res, 1999 Mar 15, 27 (6): 1517 - 1523.

[8] NCBI Resource Coordinators. Database resources of the National Center for Biotechnology Information [J]. Nucleic Acids Research, 2018, 46 (D1): D8 - D13.

[9] Zhang Z, Schwartz S, Wagner L, et al. A greedy algorithm for aligning DNA sequences. [J]. Journal of Computational Biology, 2000, 7 (2): 203 - 214.

[10] Morgulis A, Coulouris G, Raytselis Y, et al. Database indexing for production MegaBLAST searches [J]. Bioinformatics, 2008, 24 (16): 1757 - 1764.

[11] Johansson MU, Zoete V, MichielinO, et al. Defining and searching for structural motifs using DeepView/Swiss - PdbViewer [J]. BMC Bioinformatics, 2012, 13: 173.